PIPE DRAFTING

PIPE DRAFTING

WILLIAM HARTMAN

Coordinator of Curriculum, Building and Planning
Adult Services, Research
Tulsa County Area Vocational-Technical School

FRED WILLIAMS

Pipe Designer and Drafter

GREGG DIVISION
McGRAW-HILL BOOK COMPANY

New York   Atlanta   Dallas   St. Louis   San Francisco   Auckland
Bogotá   Guatemala   Hamburg   Johannesburg   Lisbon   London
Madrid   Mexico   Montreal   New Delhi   Panama   Paris
San Juan   São Paulo   Singapore   Sydney   Tokyo   Toronto

Sponsoring Editor: MYRNA BRESKIN
Editing Supervisor: EVELYN BELOV
Design Supervisor: HOWARD BROTMAN
Production Supervisor: LOU DeMAGLIE
Art Supervisor: GEORGE T. RESCH

Technical Studio: FINE LINE INC.

Library of Congress Cataloging in Publication Data
Hartman, William, (date)
Pipe drafting.

Includes index.
1.   Pipe lines—Drawings. 2.   Plumbing drafting.
I.   Williams, Fred, (date) joint author.
II.   Title.
TJ930.H39   621.8'672'0221   80-17652
ISBN 0-07-026945-9

PIPE DRAFTING

1 2 3 4 5 6 7 8 9 0   S M S M   8 9 8 7 6 5 4 3 2 1

ISBN 0-07-026945-9

CONTENTS

PREFACE

Pipe drafting has a unique language with many terms, symbols, and abbreviations that must be mastered by students who intend to become pipe drafters. This text, Pipe Drafting, not only introduces students to the basics of pipe drafting but serves as a reference book and general information source for people in a variety of pipe-related professions.

Chapter 1 introduces the subject of pipe drafting with a quick look at history and the rise of modern technology.

Chapter 2 reviews the basics of drafting. Although students using this book have probably mastered basic drafting before coming to the study of pipe drafting, a review of the basic concepts is useful at this point.

Chapter 3 discusses both piping processes and equipment and the process of developing a set of working drawings.

Chapter 4 describes pipe standards and data.

Chapter 5 discusses flanges and flange selection with particular emphasis on the standardizations of pipe sizes and applications.

Chapter 6 deals with the major types of welded flanges and threaded fittings and the particular applications of each.

Chapter 7 discusses the most common types of valves, how they operate, and what they are used for.

Chapter 8 describes the mathematics needed by the pipe drafter and acquaints the student with the various types of reference tables, charts, and conversion factors that they will need in their future profession.

Chapter 9 deals with plan elevations and fabrication drawings. By the end of Chapter 9 the student should be able to visualize views and supply missing views as necessary.

Chapter 10 discusses isometric drawings. By the end of that chapter the student should be able to illustrate pipe fittings and valves in isometric forms.

Chapter 11 is about pipe fabrication and spooling. It should enable the student to understand the basics of pipe fabrication and the implementation of spool drawings.

Chapter 12 discusses specifications in working drawings and design practices. This chapter familiarizes the student with what is required for a complete set of working drawings.

Appendixes A and B provide a set of plans and a typical list of specifications to accompany them.

Pipe Drafting is a text with a strong emphasis on the behavioral objective approach to instruction. It provides a variety of drawings, reference materials, terms, abbreviations, suggested activities, and tests to facilitate the learning process. The student, instructor, or professional drafter should find the material to the point—and very applicable to a variety of pipe-related fields.

William Hartman

Fred Williams

One

Introduction to Pipe Drafting

The principle of conducting a liquid through tubes or pipe is as old as life itself. The human body, for example, is a very intricate system of arteries, tubes, and passageways. This perhaps gave the ancient Chinese the idea of transporting liquids through hollow bamboo logs. Water was first transported to public fountains by the Romans, and later in America the pioneers bored holes in logs to accomplish the same purpose.

Pipe is made today of cast iron, steel, cement, copper, glass, plastic, aluminum, and a variety of other materials depending on the particular application. Pipe can be made by several procedures, such as molding, casting, welding, forging, drawing, and pushing a solid shaft over a mandrel to make a hole through its center. Whatever the application may be, there is a type of pipe that is best suited for it. The drafter has the responsibility of selecting the pipe that will best suit the needs of the situation (see Fig. 1-1).

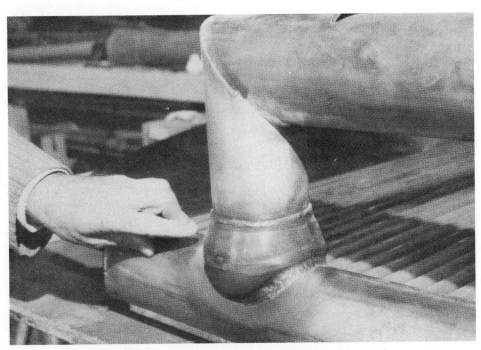

Fig. 1-1 Selecting the pipe that best suits the needs of the situation is the drafter's responsibility. *Courtesy Energy Products Corp.*

Fig. 1-2 Petroleum processing is only one application of piping.

OUR RAPIDLY ADVANCING TECHNOLOGY

The past two decades have brought about tremendous change in communication, transportation, space travel, manufacturing, life-styles, building methods, chemical technology, and medicine. The list could go on almost indefinitely. With the rapidly growing number of people on this planet we know as Earth, it will be a continuous challenge to meet the basic needs of society.

Like our highways and railroads, pipe is a means of transporting useful material. The process of making liquids from solids such as coal has made it possible to transport even solid materials through pipe. Current trends of human transportation point toward a system of large pipes and tubes that will transport capsules through them from one point to another.

We normally think of pipe only in relation to petroleum processing plants in which crude oil is refined into hundreds of by-products (see Fig. 1-2). However, pipe has many other applications that will be mentioned briefly in order to help create an awareness of the tremendous need for pipe drafting.

Each day the average home uses about a ton of water. Huge pipes bring this water to the community from wells, lakes, or other sources of supply. A vast network of pipes then distributes the water to every home. The waste system of sewer pipes that transports thousands of gallons of waste each day is also complex. Modern buildings are made of heat systems, cooling systems, waste systems, water systems, and intricate systems of conduit tubing for

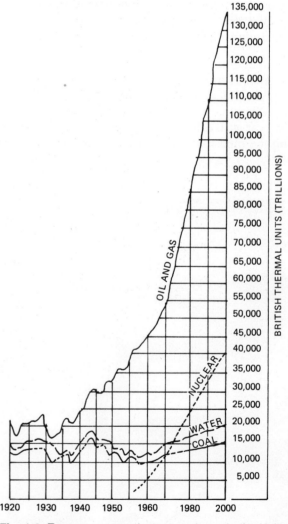

Fig. 1-3 Energy consumption comparisons in the United States.

2

wiring, all of which are prime users of pipe. Projected energy consumption in the United States is heavily dependent on oil and gas (Fig. 1-3). But all consumption methods use elaborate pipe systems.

Probably one of the most overlooked applications for pipe is the area of food preparation plants. Pipe used in these plants is usually made from stainless steel, glass, or fiberglass, all of which are corrosion-resistant materials that aid in the quality control of canned foods.

Thousands of gallons of beverages and alcohol are consumed each month all over the world, necessitating the operation of huge process plants and distilleries. They in turn require very complex systems of pipe equipment, large furnaces, boilers, and pumps. This typifies the many advances that are being made in manufacturing processes. The mechanized methods of preparing, bottling, canning, packaging, and shipping involve very complex systems of pneumatics, hydraulics, water, and other types of pipelines.

Pipe itself connects the individual pieces of equipment in a total system. A total system such as we would find in a large sea vessel, an airplane, an auto-

mobile, a truck, or perhaps a locomotive would require numerous types of pipe and tubing to make the entire system function.

Space research during the past 15 to 20 years, though costly, has contributed a great deal to our quality life-style. Miniaturization of circuits and life-support systems has led to more compact and efficient electrical devices that use various forms of pipe.

Large petroleum companies devote millions of dollars to chemical research. Pipe is used in the production of dozens of oil by-products such as fertilizers, plastics, clothing, and insecticides (Fig. 1-4).

Our recent energy crisis has brought about a trend toward solar and nuclear energy which has opened up a great demand for pipe used in collectors and storage modules. Also, distribution outlets will perhaps be a prime user of pipe in various forms.

Without exhausting this subject to the limit, it is obvious that pipe drafting has a wide and diversified field of application besides the traditional oil-related applications. The pipe drafter has a great responsibility, as well as a tremendous opportunity to select from the many related fields that use pipe drafting.

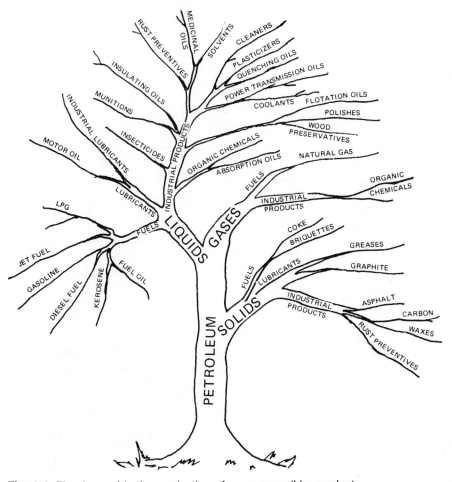

Fig. 1-4 Pipe is used in the production of numerous oil by-products.

LANGUAGE OF DRAFTING

Pipe drafting is a language in the sense that there are many terms, symbols, and abbreviations that must be mastered if one is to function effectively as a pipe drafter. Other types of drafting, such as architectural and structural, also require the student to learn the terms, symbols, and abbreviations peculiar to those fields in order to function effectively. Certain types of drafting have things in common with other types, and the drafter working in different areas begins to see how pipe drafting, for example, is separate and unique and at the same time has many things in common with some other kind of drafting. Structural, electrical, pressure vessel, and mechanical drafting are each separate types but have a very close relationship to pipe drafting. The basic principles of drafting are a universal language that can be understood in nearly all parts of the world. Competency and skill in drafting will come only with time and experience and a dedicated desire to learn this language (Fig. 1-5).

PRINTS AND REPRODUCTIONS

Pipe drafting, like any type of drafting, requires that each original drawing be reproduced by some method so that it can be preserved for future use. Original drawings are very expensive to produce and must be preserved; therefore, the drawing must be completed on some type of material that will be very durable and lend itself to much handling. The drawing material must be translucent to allow light to pass through so that the image will be transferred to the reproductive material. Four types of drawing material are used primarily: vellum, linen, Mylar, and plastic-coated film. *Reproduction* is a broad term meaning primarily the copying of an original drawing. There are four main methods used for reproduction: (1) photocopy (Fig. 1-6), (2) moist print (Fig. 1-7), (3) dry print (Fig. 1-8), and (4) blueprint (Fig. 1-9). Photocopies are rapidly becoming predominant due to quality and speed of duplication. The photocopy process is a camera method of reproduction. It can be used to enlarge or reduce the size of an original drawing.

DRAFTING TECHNIQUES

All types of drafting require a certain amount of skill and technique. The use of various drafting instruments and items of equipment makes it possible to convert ideas and theory into actual finished products. Basic drafting skill must be achieved by practice and application of conventional symbols, lines, abbreviations, notes, and theory. The drafter must strive for speed, but neatness and accuracy are

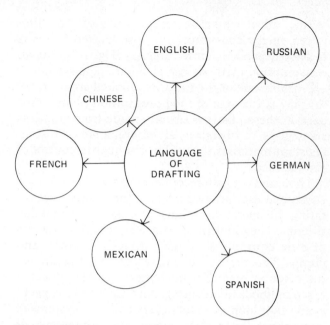

Fig. 1-5 The language of drafting is common to many cultures.

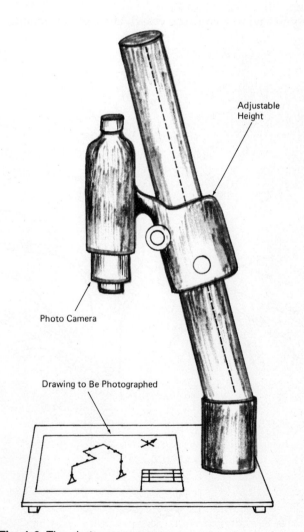

Fig. 1-6 The photocopy process.

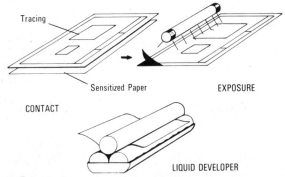

Tracing

Sensitized Paper

CONTACT

EXPOSURE

LIQUID DEVELOPER

Fig. 1-7 The moist-print process.

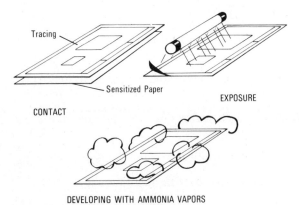

Tracing

Sensitized Paper

CONTACT

EXPOSURE

DEVELOPING WITH AMMONIA VAPORS

Fig. 1-8 The dry-print process.

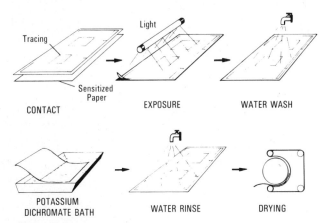

Light

Tracing

Sensitized Paper

CONTACT

EXPOSURE

WATER WASH

POTASSIUM DICHROMATE BATH

WATER RINSE

DRYING

Fig. 1-9 The blueprint process.

equally important. To achieve speed the drafter uses standard reference materials, templates, stencils, tables, and other timesaving devices like electric erasers, drafting machines, and mechanical lead pointers.

It is nearly impossible to list and describe the uses of all the equipment available, since equipment changes continuously. Calculator design can change each month, and each new model requires a separate set of instructions. As a drafter, you must learn to keep up with the current trends in equipment and procedure.

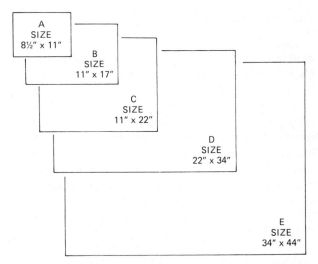

A
SIZE
8½" x 11"

B
SIZE
11" x 17"

C
SIZE
11" x 22"

D
SIZE
22" x 34"

E
SIZE
34" x 44"

Fig. 1-10 Drawings are prepared on standard paper sizes.

The drafter's overall job description is rapidly changing due to the expanded use of the electronic computer. Computer drafting enables the drafter to use preestablished programs and tapes to increase the quantity of work performed. Pipe drafting is only one type of drafting that is using computer graphics. Computers will not eliminate drafters but will change their methods for arriving at a completed working drawing. The drafter's job will be more technical and less artistic.

Whatever method is used for achieving a completed drawing, the drafter should be conscientious about the following items:

A. Line weight and contrast

B. Lettering

C. Orientation of information

D. Accuracy

E. Neatness

DRAWING COMPLETION PROCEDURE

Every company has its own system for drawing completion and its own requirements for what information should be on the completed drawing. Drawings are given numbers for identification and can be grouped into sizes, such as A (8½ × 11 in), B (11 × 17 in), C (17 × 22 in), D (22 × 34 in), and E (34 × 48 in). Drawings also are prepared in larger sizes, on rolls 34 in wide by 6 ft, 8 ft, and 10 ft long (see Fig. 1-10.)

The steps involved in the routing procedure for a drawing is very involved. Drawings should originate from a designated stage of the process and systematically flow through each phase. The orderly flow ensures quality control and eliminates confusion, misunderstandings, and costly errors.

CONCLUSION

This chapter has included a brief history of pipe drafting. Some of the present-day applications and changes that are taking place have also been mentioned to help the student see the impact pipe drafting actually has on society. The tremendous demand for pipe systems has created a very lucrative opportunity in the field of pipe drafting. Pipe drafting seems very complex at first, but with determination and effort a career in pipe drafting can be very well paid with tremendous opportunities for advancement.

EXERCISES

1-1. Define the following terms:

a. Blueprint

b. Language of drafting

c. Drawing completion sequence

d. Reproduction

e. Drafting technique

1-2. Name five applications for pipe drafting not mentioned in this text, and research their use in the field.

1-3. Make a list of as many items of equipment used in pipe drafting as you can find in your drafting setting.

1-4. Use reference material such as an encyclopedia or a drafting text to name and briefly describe the function of each process.

1-5. Visit a drafting department and have the chief drafter lay out a routing schedule for drawing completion.

JOB ASSIGNMENT

1-6. On an 8½ × 11-in sheet of vellum make a tracing of a simple piping drawing from this text. Use this tracing to run a print, and critique the reproducibility of the print.

REVIEW QUESTIONS

1-7. What liquid was first transported using a form of pipe?

1-8. Name five types of material used to make pipe today.

1-9. How much water does the average home use per day?

1-10. Name ten applications for modern-day piping.

1-11. Explain how drafting is a language.

1-12. Explain why prints are necessary.

1-13. Write a brief explanation of the importance of drafting technique.

1-14. Create your own routing procedure for a typical drafting job.

Two

Basic Drafting Review

Although a student should have had basic drafting prior to entering a study of pipe drafting, it is often beneficial to review basic ideas and conventional practices. This chapter will deal with a few of the basic concepts that are common to many areas of drafting.

SCALE USAGE

The term *scale* usually denotes an instrument used in making measurements (Fig. 2-1). However, this term also means to enlarge or reduce the size of an object on a drawing so that it can fit on a specific size of paper and be dimensioned adequately. Notice the comparative size of the same object at different scales in Fig. 2-2.

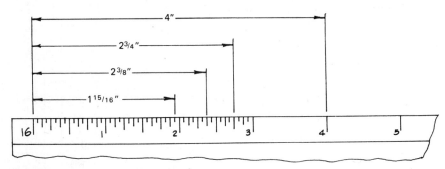

Fig. 2-1 Measurements at scale of 12″ = 1′−0″.

There are several types and shapes of scale instruments. Each instrument has a specific application. Some of the basic shapes are (a) triangular, (b) opposite bevel, and (c) flat bevel (see Fig. 2-3, p. 8). The most common shape in piping is the triangular scale.

In architectural, piping, structural, and pressure vessel drafting, the *architects' scale* is most commonly used because it is graduated in feet and inches, which are the common units of measurement (see Fig. 2-4). The eight scale ratios on the architects' scale are listed at right:

$^3/_{32}$ in	= 1 ft 0 in	1:128 ratio
$^1/_8$ in	= 1 ft 0 in	1:96 ratio
$^3/_{16}$ in	= 1 ft 0 in	1:64 ratio
$^1/_4$ in	= 1 ft 0 in	1:48 ratio
$^3/_8$ in	= 1 ft 0 in	1:32 ratio
$^1/_2$ in	= 1 ft 0 in	1:24 ratio
$^3/_4$ in	= 1 ft 0 in	1:16 ratio
1 in	= 1 ft 0 in	1:12 ratio

On plan and elevation views of piping systems, the most common scale is ⅜ in = 1 ft 0 in (Fig. 2-5). This scale means that ⅜ of an inch is graduated into 12 parts, and each of those parts represents an inch. A scale of ⅜ in = 1 ft 0 in is at a ratio of 1:32.

Practice reading the ⅜-in scale ratio and transfer this skill to the other ratios. The principle and the procedure are the same and can be learned very easily. Study the ⅜ in=1 ft 0 in scale in Fig. 2-6, and

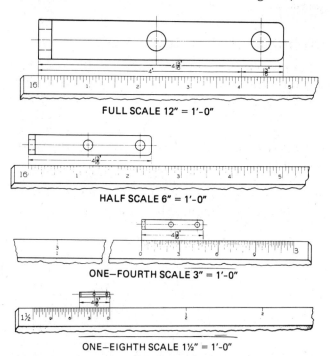

FULL SCALE 12" = 1'-0"

HALF SCALE 6" = 1'-0"

ONE—FOURTH SCALE 3" = 1'-0"

ONE—EIGHTH SCALE 1½" = 1'-0"

Fig. 2-2 Comparative scale sizes of same object.

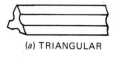

(a) TRIANGULAR

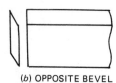

(b) OPPOSITE BEVEL

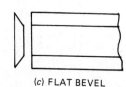

(c) FLAT BEVEL

Fig. 2-3 There are several shapes of scale instruments. Triangular, opposite bevel, flat bevel are a few of the basic shapes.

practice reading the scale using the exercises at the end of this chapter.

DIMENSIONS, NOTES, AND SPECIFICATIONS

A working drawing actually does two basic things: It shows the shape of an object by views, and it shows the size of the object by dimensions, notes, and specifications. The drafter must learn not only the mechanics of drawing the views of the object but also the correct way to dimension, the proper way to state notes, and how to conform to given specifications—or perhaps even to write the specifications under certain conditions.

Dimensions: Dimensions are usually expressed in feet and inches and are placed above dimension lines (see Fig. 2-7). Dimensions, in the form of elevations and coordinates, are also placed on the drawing to locate centerlines, etc., in relation to north, south, and groundlines.

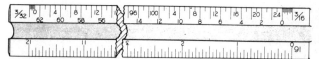

Fig. 2-4 The architect's scale.

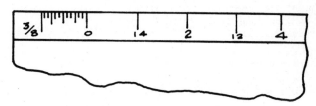

Fig. 2-5 On plan and elevation drawings, the most common scale is ⅜" = 1'-0" scale.

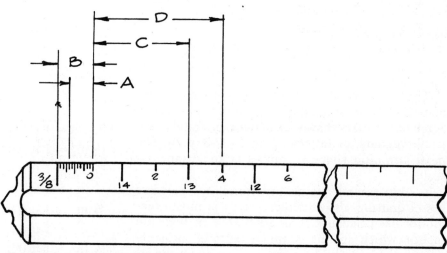

Fig. 2-6 Reading the ⅜" = 1'-0" scale.

Notes: There are two basic kinds of notes, general and specific. *General notes* are placed in a numbered note column (see Fig. 2-8) and apply to the whole drawing. An example would be a weld description that applied to all welds. However, a *specific note* takes precedence over a general note; it is attached directly to a specific item on the face of the drawing with a leader line, as in Fig. 2-9.

Specifications: "Specs," as they are often called, are mostly word descriptions based on customer requirements of minimum or acceptable material and procedure standards.

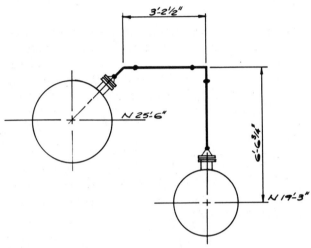

Fig. 2-7 Dimensions on pipe drawings are usually expressed in feet (') and inches (").

General Notes:

1. ALL 90° ELBOWS TO BE LONG RADIUS.
2. ALL FLANGES 150# RAISED FACE.

Fig. 2-8 General notes apply to the whole drawing.

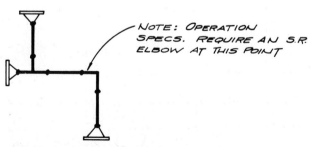

NOTE: OPERATION SPECS. REQUIRE AN S.R. ELBOW AT THIS POINT

Fig. 2-9 Specific notes are attached directly to the specific item.

BASIC TYPES OF LINES

Lines are used to represent the shape of an object as well as to indicate its size. The alphabet of lines is a listing of the basic lines and their individual features. The pipe drafter should be familiar with all the types of lines and should learn how they apply to each situation. Learning the use of lines is much like studying a group of symbols or perhaps a new language.

ORTHOGRAPHIC PROJECTION

Orthographic projection is the basis by which the relative positions of objects are known. In all types of drafting, the basics of orthographic projection apply. *Orthographic projection*, properly defined, is the method of representing the exact form of an object in two or more views on planes at right angles to each other.

Views are projected to an imaginary picture plane box and then the picture plane box is laid flat, as on a piece of paper. The principle can be seen clearly in Figs. 2-10 and 2-11.

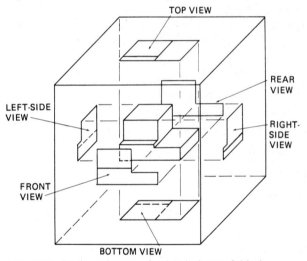

Fig. 2-10 Picture plane projection being unfolded.

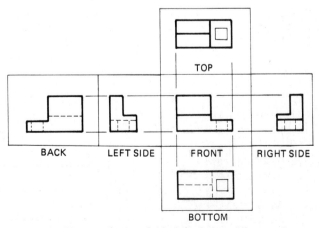

Fig. 2-11 Picture plane unfolded depicting different views.

Figure 2-12 shows the correct relationship of views after they have been laid out flat. The principle is the same in piping, machine, and other kinds of drafting.

Basic drafting books use blocks for illustration because the manner of rotation can be seen easily. The drafter must learn to visually grab the object, rotate it, and then imagine the view needed. Use Fig. 2-12 to practice this rotation principle in your mind. Later this principle will be related to pipe, and it is imperative that it be clear in your mind.

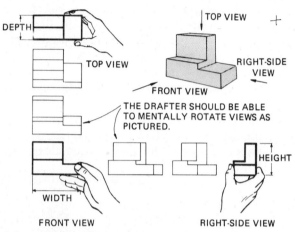

Fig. 2-12 Practice the rotation principle in your mind. Drafters should be able to mentally rotate the views as pictured.

SECTIONAL VIEWS

Sectional views are an invaluable tool to the pipe drafter. Sections are used to cut through a particular area so that the internal parts can be shown and dimensioned clearly. In machine drawing there are several types of section views—full, half, removed, broken, etc.—but in piping the removed full section is used most commonly. The full section cuts completely through an area and is indicated by a cutting plane line as shown in Figs. 2-13 and 2-14. The principles of orthographic projection apply to sectional views.

In Fig. 2-13, B indicates the section view and 3300–D–4090 indicates the sheet number that the section is taken from.

A thorough understanding of section views should be one of the most valuable tools of the pipe drafter (see Fig. 2-15).

PICTORIAL VIEWS

As in all areas of drafting, it is often necessary to show a pipe layout so that it looks realistic or like a picture. This is called a *pictorial view*, and there are

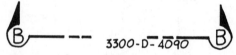

Fig. 2-13 Cutting plane pipe notation.

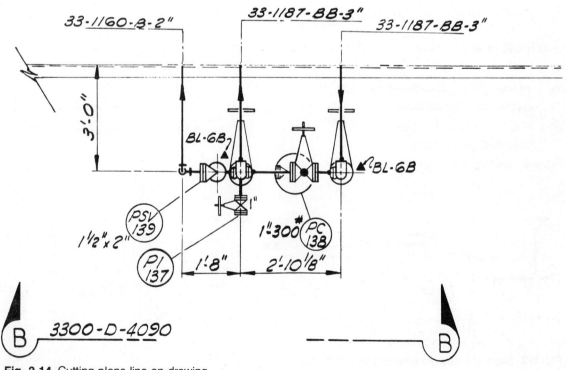

Fig. 2-14 Cutting plane line on drawing.

several types of pictorials, but the type used most often in piping is the isometric view.

An *isometric* shows all planes of an object on an equal axis and is very useful in illustrating all the angles, turns, and directions that pipe may conform to. The drafter must determine what position of the isometric view best represents the pipe layout.

Isometric Axis: An isometric drawing is laid out using the North-South axis lines to represent the direction of pipe runs (see Fig. 2-16). Pipe is drawn in isometric to conform to its actual flow direction.

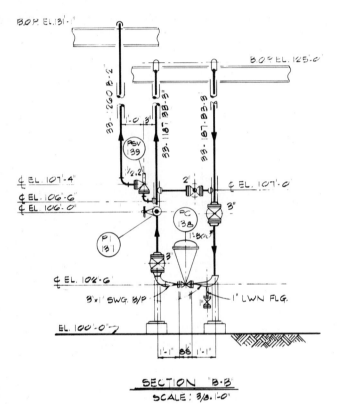

Fig. 2-15 Partial pipe section from Fig. 2-14.

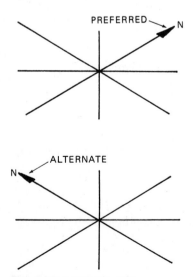

Fig. 2-16 Isometric axis.

Isometric Dimensions: Dimensions are placed on isometrics so that the correct dimensional relationship can be expressed. Isometrics in piping are *not* drawn to scale but are expanded or condensed in certain areas to best show fittings, valves, etc., so that these parts do not block each other out of view (see Fig. 2-17.)

Isometric Angles: It would be relatively easy to dimension pipe drawings if pipe all ran at perpendicular angles, but this is not the case. Pipe is sloped, skewed, and curved many times, and these turns must be shown in the form of dimensions and notes. Angles are not drawn with the protractor on an isometric drawing, but are drawn with offsets (see Fig. 2-18). Keep in mind that isometrics are not drawn to scale. Isometrics are covered in greater detail in Chap. 10.

AUXILIARY VIEWS

Auxiliary views are additional views drawn of an area that is slanted or sloped in the normal views. In an auxiliary view the sloped surface can be shown in its normal shape (see Fig. 2-19).

With an auxiliary view pipe can be shown in a normal plane for ease in dimensioning. In pipe drafting an auxiliary view may need to be projected from any direction, from any view. A good basic understanding of auxiliary-view projection will be most beneficial in later pipe drafting assignments.

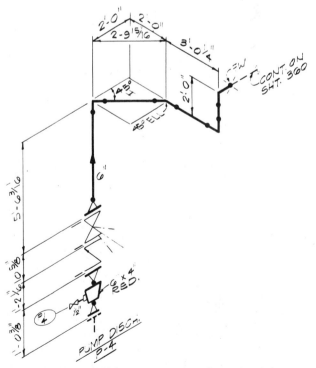

Fig. 2-17 Isometric drawings are not drawn to scale.

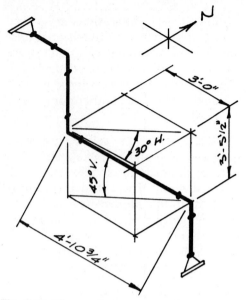

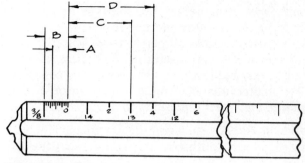

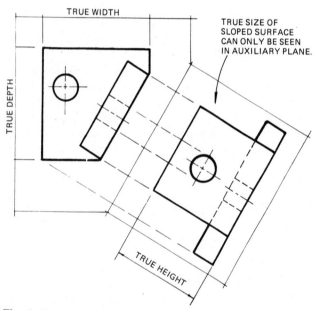

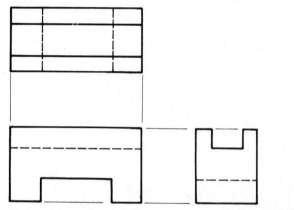

Fig. 2-18 Pipe angles represented in isometric.

Fig. 2-19 An auxiliary view of a sloped surface.

Fig. 2-20 Scale reading.

Fig. 2-21 Isometric sketch.

c. Size description   d. Specific notes
e. Orthographic projection   f. Section view
g. Cutting plane line   h. Auxiliary view
i. Axis line   j. Specifications
k. Isometric view

2-2. Match the terms and abbreviations:

a. Projection	1. pl.
b. Specifications	2. ortho.
c. Centerline	3. aux.
d. Auxiliary	4. proj.
e. Dimensions	5. dim.
f. Orthographic	6. iso.
g. Isometric	7. specs.
h. Plane	8. C_L

2-3. On a separate sheet of paper, number from A to D and read the dimensions shown in Fig. 2-20.

2-4. Make an isometric sketch of the three views in Fig. 2-21.

2-5. Match the isometric views with the correct orthographic view. Number your paper from 1 through 4 and put the letter beside the number it matches (see Fig. 2-22).

2-6. Complete the front view as a section view (see Fig. 2-23).

CONCLUSION

Study the basic principles of drafting, and they will make further study in pipe drafting much easier and more meaningful. If more study in depth is needed in this area, study a good basic drafting text and become more familiar with basic theory and practices.

EXERCISES

2-1. Define the following terms:

a. Scale   b. Shape description

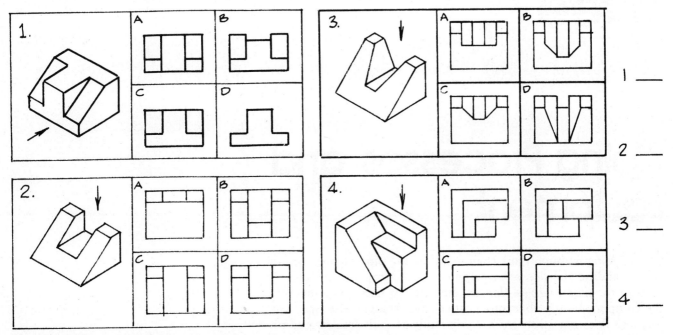

Fig. 2-22 Orthographic and isometric views.

1 ___

2 ___

3 ___

4 ___

Fig. 2-23 Sectional view.

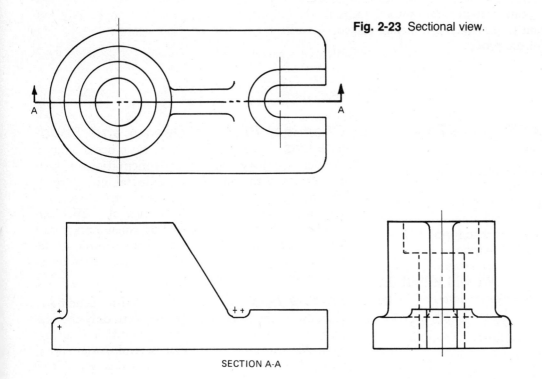

SECTION A-A

Three

Piping Processes and Equipment

Pipe drafting is much more than just drawing pipe; it involves an understanding of *processes, equipment, structure,* and *utilities,* plus all the *pipes, valves, flanges,* and *fittings* used in connecting the components together into a functional arrangement. Space in this text does not allow an in-depth study of each area related to pipe drafting, but a brief description of each should clarify pipe drafting and make its application more meaningful. The process of developing a set of working drawings follows a definite sequence, and it is important to understand the total project as well as each particular stage of the process.

STAGES OF DEVELOPING A SET OF WORKING DRAWINGS

The conception of a process plant begins when a chemist develops a new or improved method of refining a product. From experimentation the chemist concludes a sequence of chemical stages that a raw product has to go through in order to form a refined product that can be put on the market.

From this basic start there are several steps of drawing completion necessary. We will list all the steps and then discuss each in more detail as we go along.

A. Chemical process design and specifications

B. Process flow sheet

C. Mechanical flow diagram

D. Plot plan layout

E. Plan and elevation layout

F. Scale model

G. Isometric line (erection drawings)

H. Spool sheet

(It should be noted here that the above sequence of items is flexible and at times some drawings are combined or eliminated as the project changes.)

Chemical Process Design and Specifications: A particular chemical process is often derived after long hours of experimentation and testing. At times the process is accidental, but more often than not it is the result of intentional patterns of research: The chemist tries to come up with a particular product that is needed by consumers. Since the product must conform to certain criteria, its development must follow certain specifications as guidelines (see Fig. 3-1).

Process Flow Sheet: A single-line schematic that shows the basic process flow with only the major equipment is called a *process flow sheet* (see Fig. 3-2). All the pipe sizes, valves, and dimensional valves have not been determined at this point. The process flow sheet will be used to develop all the stages that follow.

Mechanical Flow Sheet: Also a single-line schematic, the mechanical flow sheet takes the process flow sheet a step beyond and includes major equipment items, instrumentation, major valves, and line sizes (see Fig. 3-3). The mechanical flow sheet is used with the plot plan to develop an actual set of plan views and sectional views to a definite ⅜ in = 1 ft 0 in scale.

14

Fig. 3-1 Chemists are constantly trying to develop new and improved methods of product refinement.

Plot Plan: Plot plans are started from field information for the purpose of laying out and locating facilities on a particular plot of ground. Scales will vary in order to make the sites fit the drawing sheets. The shape of the plot is not always rectangular but is often unusual. A gridded coordinate system is used to locate equipment. The grid system is usually set up in such a way that coordinates can be read in two directions only. The zero-zero coordinate appears at the extreme north or south and at the extreme east or west edge of the plan site. Zero-zero should not be on a fence line or on any line that will be obstructed by a plant facility as construction progresses.

The grid system should indicate coordinates on all main pipe racks, building lines, centerlines of roads, and other such places as deemed necessary by the engineer. Where coordinates of building lines or building centerlines are shown, these are noted on the plot plan.

After the grid system is established and locations of facilities are set, care must be exercised in the location of plan areas to avoid splitting main facilities into two or more plan drawings (see Fig. 3-4, p. 17).

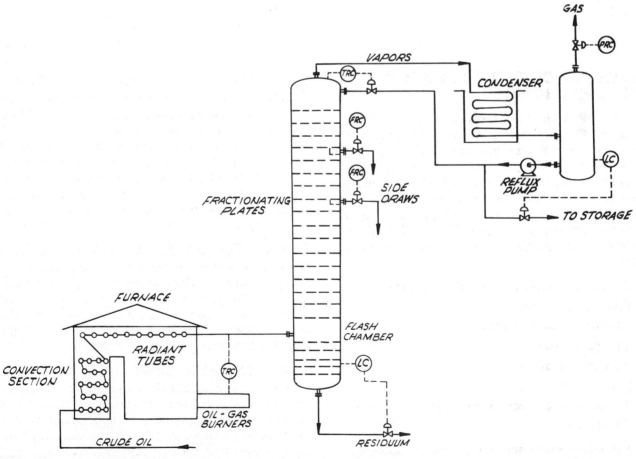

Fig. 3-2 A process flow sheet is a single-line schematic that shows the basic process flow and the major equipment.

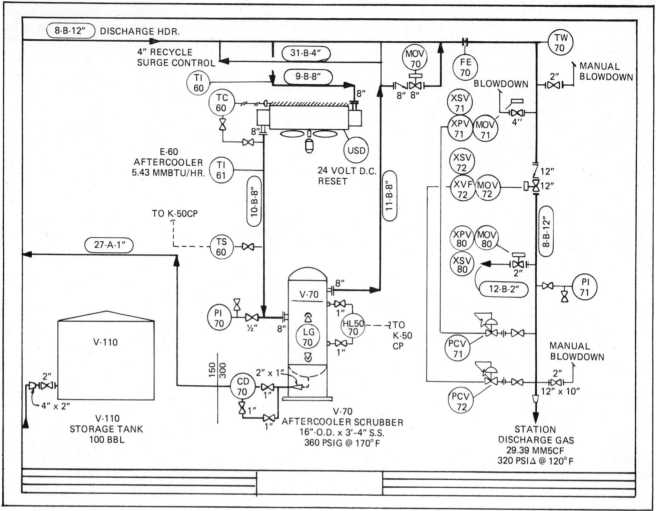

Fig. 3-3 A mechanical flow sheet is a single-line schematic that includes the process flow, major equipment, instrumentation, major values, and line sizes.

Plans, Sections, and Elevations: The piping plan areas are laid out and numbered consecutively on the plot plan prior to the start of piping plans. Drawing numbers of piping plan drawings should be indicated in the areas shown on the plot plans. Match lines on piping drawings are noted. *Match lines* are the boundaries of sheets 1 that match those of other sheets, so that lines can continue from one sheet to another. Coordinates must be shown on match lines and piping of facilities must be dimensioned once in each of four directions to tie in to match lines. The dimensions tying the equipment or facilities to the match line should be located in such a way that when a finished drawing is made from the individual drawings, the checker has dimensions for the overall dimensional check of the layout.

Plans, elevations, and sections are usually drawn to the scale ⅜ in = 1 ft 0 in. They also are dimensioned and show line size, specification number, valve types, flow direction, coordinate locations, match lines, equipment identification, general notes, material callouts, sectional-view locations, and a great deal of other information. (Refer to the plan view in Appendix A, p. 158.) Only basic overall dimensions are used, necessitating the additional use of individual line drawings (isometrics), which go into more depth in showing dimensions and details (see Fig. 3-5, p. 17).

Scale Model: The labor involved in making one is very costly, but a scale model is helpful during the erection stage of a facility. Models make the process of detailing isometric spools and plan sheets more visual, thus cutting down on costly mistakes. Some jobs lend themselves very well to the use of scale models, but on others they are not feasible. Scale models are eye-catching and interesting to look at, as well as very useful design models that often save engineering time and prevent erection mistakes (see Fig. 3-6, p. 18).

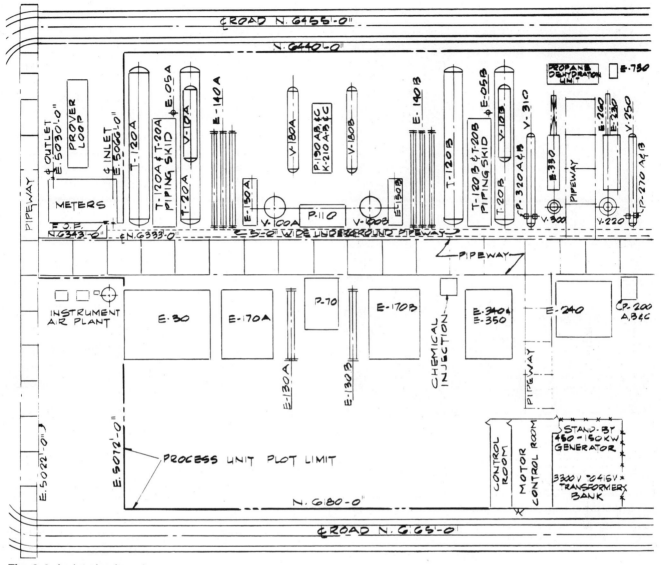

Fig. 3-4 A plot plan layout.

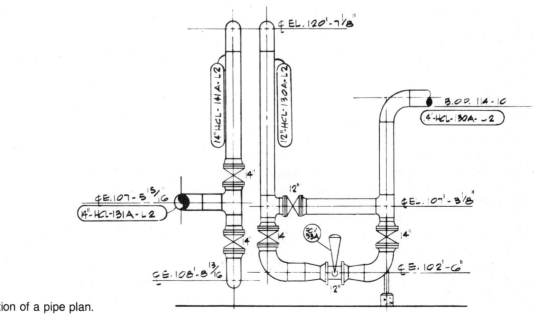

Fig. 3-5 Detailed section of a pipe plan.

Fig. 3-6 A scale model of a plant facility.

Isometric Lines: As was mentioned previously regarding plans and elevations, it is necessary to have isometric drawings of each pipeline so that each line can be completely dimensioned individually, apart from all the clutter of other information found on a plan and elevation drawing. The isometric is not to scale but is simply a schematic showing all the components, valves, pipe, fittings, and overall dimensions peculiar to that line (see Fig. 3-7, p. 19). Even though it is not to scale, the correct scale dimensions are placed in their correct locations.

Spools: Drawings called *spools* divide the sometimes very long pipe isometric into lengths and sizes that can be fabricated in a weld shop and shipped to a job location. Spools usually break at flange or valve connections. A spool includes a bill of material and all dimensions and notes necessary to put together a run of pipe (see Fig. 3-8, p. 19). A pipe spool is the final result of all earlier drawings. Without individual spools that will actually fit together, all the planning and preparation are of no importance.

GENERAL PROCESS-PLANT TERMS

There are several general process-plant terms that relate to crude-oil refineries and chemical plants. A basic understanding of each would be most beneficial as the student begins to learn this language of pipe drafting.

Tank Farm: Reserve and product tanks, sometimes 200 ft in diameter, usually away from the main plant, are used much like the fuel tank on an automobile (see Fig. 3-9, p. 20).

Gasoline Plant: A gasoline plant uses gas vapor from the crude-oil refining process and breaks it down into lighter gases (such as methane) and heavier gas by-products (such as gasolines). A gasoline plant is just a part of a total oil refinery.

Hydrocarbon: Hydrocarbons are combinations of hydrogen and carbon, and form the base for all petroleum products.

Refinery: This is a plant that refines crude oil into many dozens of oil and gas by-products such as jet fuel, kerosene, fuel oil, and asphalt (see Fig. 3-10, p. 20).

Chemical Plant: The basic ingredient in a chemical plant is crude oil or gas products, but the finished products are a variety of many chemicals, medicines, cleaners, etc.

Underground Storage: Underground caves (natural or constructed) are used to store charge stock as well as finished products.

BASIC EQUIPMENT TERMS

Equipment items are the vessels, pumps, etc., that are required step by step to refine or aid in refining the raw charge stock to the finished product. Some of the major equipment items are listed and briefly defined below:

Pressure Vessel: There are several types of pressure vessels used in process pipeline work. Basically a *pressure vessel* is a vessel of some type that operates at a pressure greater or less than its surrounding atmosphere. Vessels come in three basic shapes: vertical cylinder, horizontal cylinder, and sphere. In addition, there are combinations of these shapes (see Fig. 3-11, p. 21).

As for function, there are two main categories of pressure vessels: (1) drums and tanks, which do not have internal accessories such as accumulators, receivers, knockouts, or storage tanks; and (2) towers, which include fractionating towers, reactors, regenerators, etc. Towers have internal accessories and are often very complicated (see Fig. 3-12, p. 22).

Heat Exchanger: A heat exchanger simply exchanges heat from one stream to a second stream. The streams do not intermingle, but the heat is transferred by conduction through the walls of pipes

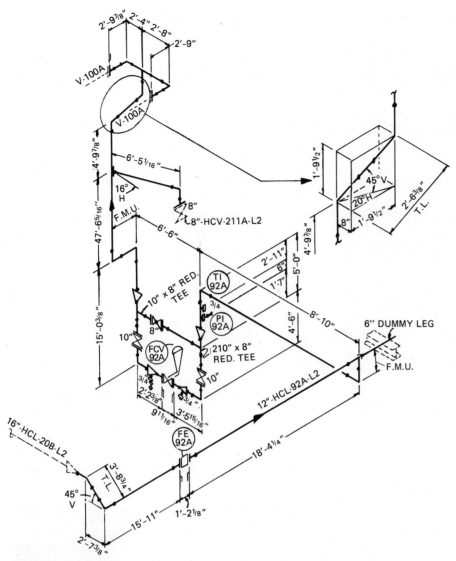

Fig. 3-7 An isometric pipe plan.

MK	QUAN.	DESCRIPTION
A	2	4 STD. L.R. 90° ELL
	2	4" END PROTECTOR
		STD. A-53-B SMLS.
B	1	4" DIA. 11'-1¼" LG. 2BE

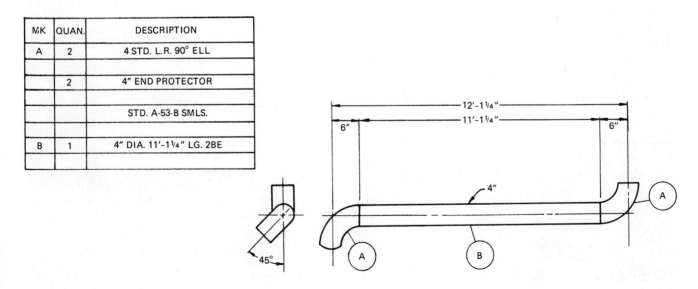

Fig. 3-8 A spool includes a bill of materials and all necessary notes and dimensions.

Fig. 3-9 Tank farms are usually located away from the main plant. They act as a storage depot for refineries and chemical plants. *Courtesy Maloney Crawford Tank Corp., Tulsa, Oklahoma.*

Fig. 3-10 Refineries refine crude oil into a number of oil and gas by-products.

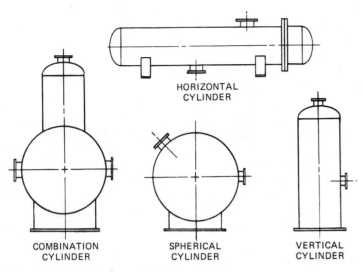

HORIZONTAL
CYLINDER

COMBINATION
CYLINDER

SPHERICAL
CYLINDER

VERTICAL
CYLINDER

Fig. 3-11 Basic shapes of pressure vessels.

or tubes. Heat exchangers are usually the shell-and-tube type or the fin-and-fan type. The latter uses air to circulate around coils as in a car radiator (see Fig. 3-13, p. 22). A very simple example is a car air conditioner.

Feedwater Heater: This is a type of vessel used to heat water to steam, transfer the heated steam to an exchanger unit, condense the steam back to a liquid, and recycle it through the system.

Compressor: A compressor increases the pressure of a vapor by reciprocating or centrifugal motion. Air, gas, and vapors can be compressed much the same as air is compressed in a balloon. Compressing a vapor increases its head pressure (see Fig. 3-14, p. 22).

Pump: Reciprocating and centrifugal pumps are used to increase flow and relay pressure as a liquid moves along a system. Without pumps, liquids would be inert except for compressed vapors upstream that would move the liquids ahead (see Fig. 3-15, p. 23).

Fired Heater: Fired heaters are vertical or horizontal units heated with burners at sides or bottom. The heaters have internal accessories such as tubes and pipes, which are located in the *radiant section.* These tubes and pipes absorb heat from the liquid around them, which has been heated by the burners. The convection section of a heater is above the radiant section and is often used to produce steam. A stack is located at the top of the unit (see Fig. 3-16, p. 24).

Fig. 3-12 Heat exchanger.

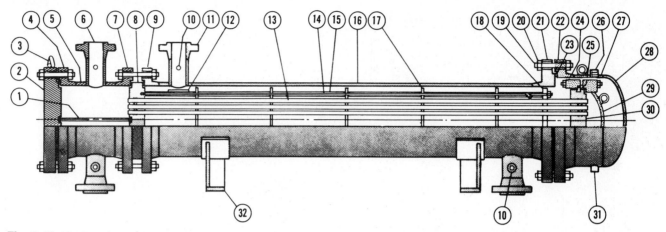

Fig. 3-13 Heat exchangers exchange heat from one stream to a second stream.

1. Pass partition
2. Blind flange
3. Lifting ring
4. Channel flange
5. Channel cylinder
6. Channel nozzle
7. Channel flg. (shell end)
8. Stationary tube sheet

9. Shell flg. (Chan. end)
10. Instrument connection
11. Shell nozzle
12. Impingement baffle
13. Tube
14. Tie rod
15. Spacer
16. Shell cylinder

17. Traverse baffle
18. Support plate
19. Stud
20. Hex nut
21. Shell flange (cover end)
22. Shell cover flange
23. Gasket
24. Back-up ring

25. Split-key ring
26. Vent connection
27. Shell cover cylinder
28. Shell cover head
29. Floating head cover
30. Floating tube sheet
31. Drain connection
32. Support saddles

Fig. 3-14 Compressors increase the pressure of a vapor. *Courtesy Alloy Compressors Limited, Glasgow, Scotland.*

Fig. 3-15 Reciprocating pumps: (*top*) drum pump; (*bottom*) reflex pump. *Courtesy Goulds Pumps, Seneca Falls, New York.*

Boiler: A boiler is another type of heater used primarily for generating steam (see Fig. 3-17, p. 24).

UTILITY TERMS

Utilities include plant items such as fuel, electricity, drains, etc. These utilities keep the plant process going even though they are not a part of the process. Utilities in a plant are much the same as utilities in a home. Some of the basic types are described briefly below:

Fuel Oil: Fuel oil is a refinery product that is used to heat homes and industry but is also used as a fuel in the refinery to keep the plant in operation.

Steam: Steam is produced by heat. It is invisible. Steam is used in many parts of the plant. Condensate is heated and converted to pure water and then further heated and converted into steam.

Condensate: As steam is used and relieved of its heat, a liquid called *condensate* is produced. The condensate is heated to a steam again and recycled. This process is repeated over and over again.

Cooling Water: Water begins at a cooling tower, circulates through the system, and returns to the tower for cooling. Cooling water is a very important element in the cool operation of a plant.

Instrument Air: A highly complex system of piping is used to direct instrument air to various valves and controls throughout the plant. Instrument air is under pressure and operates from a central control tower.

Fig. 3-16 Fired heater.

Fig. 3-17 Boilers are primarily used to generate steam.

Utility Air: Utility air is compressed air that is used to drive motors, clean out lines, blow material off objects, and dry certain areas.

Drain Outlets: Most plants have an underground system of drains and a sloped-surface system on top of the ground that collects any runoff and carries it away. Most plants have at least three runoff systems: a storm sewer, an oily water system, and an acid or chemical system. Depending on the type of plant, it may have several other systems.

Flare Burn-off: , Excess gas often builds up in equipment and is piped off to a flare stack. Here the gas burns and prevents air pollution.

PIPE FITTINGS AND FLANGES

A *pipe fitting* or *flange* is any type of connector that is used to join valves, pipe, and equipment to make the system function. Without fittings and flanges it would be impossible to change pipe size, control direction, or make connections at valves. Elbows, weld-neck flanges, tees, laterals, reducers, and returns are just a few of the basic types of fittings and flanges (see Fig. 3-18, p. 26). These are discussed in greater detail in Chaps. 5 and 6.

PURPOSE AND DEFINITION OF PIPE

Pipe is a means of transferring liquids or gases from one piece of equipment to another. Pipe is the supply line in a plant, just as arteries and veins are the supply lines in the human body, supplying the various organs with nutrients. Pipe is made of many different types of materials ranging from glass to cast iron. Its size varies as does its wall thickness (see Fig. 3-19, p. 27).

FUNCTION OF VALVES

A *valve* is a piece of equipment which is installed in a piping system for turning on and shutting off, or perhaps for controlling, the flow of liquids or gases. Pipe systems are designed to operate at certain temperatures and pressures; therefore it is necessary to use valves that are designed to operate within certain limits. Chapter 7 describes valves and how they work in great depth.

INSTRUMENTATION AND ITS APPLICATION

Instruments are devices that are used to control and coordinate the operation of valves. The instruments are connected by a series of hydraulic, pneumatic, and electrical lines. All these are controlled from a central control station. The operator monitors the systems and makes necessary adjustments to keep the system flowing efficiently.

EXERCISES

3-1. Write a brief description of each of the following:

a. Chemical process
b. Process flow
c. Plot plan
d. Erection drawing
e. Spool sheet
f. Isometric line layout
g. Mechanical flow
h. Scale model
i. Plan and elevation
j. Coordinate grid system
k. Match line
l. Pipe rack
m. Tank farm
n. Gasoline plant
o. Refinery
p. Hydrocarbon
q. Chemical plant
r. Pressure vessel
s. Tower
t. Heat exchanger
u. Feedwater heater
v. Pumps
w. Fitting
x. Flange
y. Valve
z. Instrumentation

3-2. Match the terms and abbreviations.

1. P. & E.	a. Mechanical
2. mech.	b. Pressure vessel
3. iso.	c. Feedwater heater
4. P.V.	d. Flanges
5. F.H.	e. Plan and elevation
6. utl.	f. Isometric
7. flgs.	g. Fittings
8. ftgs.	h. Drain
9. instr.	i. Condensate
10. dr.	j. Utility
11. cond.	k. Instrumentation

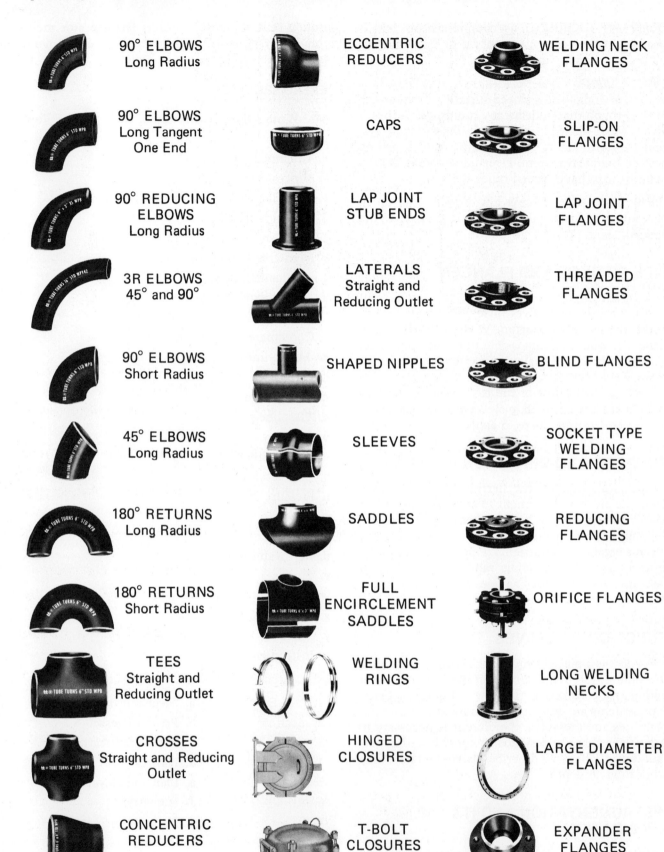

Fig. 3-18 Various types of fittings and flanges.

Fig. 3-19 Pipe comes in many sizes and materials.

3-3. Write a brief explanation of the stages involved in developing a set of working drawings.

3-4. Make a picture collection from magazines of different types of process-plant equipment.

3-5. Find the local addresses of the following valve manufacturers and either visit in person or write, requesting bulletins or pamphlets relating to pipe, valves, and fittings.

Crane

Lukenheimer

Taylor Forge

Tube Turn

Laddish

DRAWING PROBLEMS

3-6. Redraw the process flow sheet in Fig. 3-2. Draw not to scale but to correct proportion.

3-7. Redraw the plot plan in Fig. 3-4. Select the correct scale and redraw as is.

3-8. Make a block diagram depicting the stages of developing a set of plans from chemical process to spool sheets.

3-9. Instructor: Assign drawing from the set of plans in the Appendix (portion of a mechanical flow sheet).

REVIEW QUESTIONS

3-10. List the stages of drawing development in order.

3-11. How does a process flow sheet differ from a mechanical flow sheet?

3-12. How are tanks, towers, and equipment located on a plot plan?

3-13. What scale is used for most pipe plans, elevations, and sections?

3-14. Why are scale models used in plan development?

3-15. To what scale should a line isometric drawing be drawn?

3-16. Define the importance of a spool drawing.

3-17. How does a gasoline plant differ from a refinery?

3-18. What are the three basic shapes of pressure vessels?

3-19. List two types of pumps used primarily in process-plant facilities.

3-20. List five types of pipe fittings.

3-21. List three types of valves and describe basically how they work.

3-22. What is flare burn-off?

3-23. Explain basically how a heat exchanger works.

3-24. List several items of information found on a plan and elevation drawing.

Four

Pipe Standards and Data

A form of pipe was used in ancient Egypt to carry water into large cities from the water basins outside. Since that time pipe has become very sophisticated. It now has hundreds of applications in homes, industry, transportation, manufacturing, and numerous other areas. In the area of process piping, pipes are the arteries by which a process pipeline system is able to function.

The selection of pipe for a particular situation is dependent on what is going through the pipe and the pressure and temperature of the contents. By using various charts and reference guides the drafter can determine the proper application of pipe. It is our objective that the student gain a sense of respect for the complexity of pipe and its many design factors.

PIPE AND COMMON PIPE MATERIALS

Pipe is the connecting link between two or more pieces of equipment, tanks, valves, or whatever needs to be joined together for the purpose of transporting liquids, gases, or heavy-solid-suspended fluids.

Pipe is fabricated from many materials in order to satisfy the various services for which it is intended. The most common materials used for pipe fabrication are alloy, stainless steel, carbon steel, cast iron, and copper.

Alloy Pipe: There is an increase in the physical properties of alloy pipe over carbon-steel pipe which makes it more efficient for very high temperatures and pressures and also increases its corrosion-resistant value.

Stainless-steel Pipe: Stainless-steel pipe, a combination of steel, chromium, and nickel, is used for special duties. Type 304, which is also known as *18-8 stainless steel,* is 18 percent chrome and 8 percent nickel and is commonly used when stainless-steel pipe is required.

Carbon-steel Pipe: Carbon-steel pipe is adaptable to almost any service, and for that reason it is most common in a refinery. Some of the services which use carbon-steel pipe are oil, steam, water, and air.

When carbon-steel pipe is used underground or aboveground for drinking water services, it is commonly galvanized to increase its corrosion and rust resistance. *Galvanizing* is the application of a thin coating of zinc inside and outside the pipe. Carbon-steel pipe which is not galvanized is called *black pipe.*

Cast-iron Pipe: Cast-iron pipe is used where temperatures and pressure are low, such as in water or drainage service. Cast-iron pipe has a higher corrosion-resistant value than carbon-steel pipe, and for that reason it is used for underground installation of water and sewer systems. Where the corrosion factor is very high, the practice of cement lining of cast-iron pipe is employed.

Copper and Brass Pipe: Copper and brass pipe has a very high corrosion-resistant value. However, this pipe is not very commonly used because its cost is much higher than that of other material that would be just as satisfactory in most situations. Another reason why it is not a popular material in a refinery is that, due to its low melting point, it becomes a fire hazard in an emergency.

Copper Tubing: Copper tubing is very common in a refinery. It is softer and more pliable than copper and brass pipe, and it also has a high corrosion-resistant value. It is used generally for steam, air, and oil service, but most commonly for steam tracing of pipe or equipment. The steam tubing is placed next to the pipe or equipment, and then the steam tubing and the pipe or equipment are insulated together. The purpose of this service is to protect a fluid within the pipe or equipment against freezing or to keep it warm so that its viscosity is low enough to allow it to flow freely.

Copper tube sizing differs from that of steel pipe. A ¾" tubing has an outside diameter of ¾". Copper tubing comes in two types, hard and soft. Types K (hard and soft) and L (hard and soft) in sizes ⅜ to 12", and type M (hard) in sizes 2½ to 12", conform to the *American Standard for Copper Water Tube.*

Type K tube is recommended for general plumbing and heating systems, and especially for underground use and where conditions are severe. This tube is also used for gas, oil, and steam. Type L tube is recommended for interior use in general plumbing and heating work. Type M tube is used with soldered fittings only, for waste, vent, interior drainage, and other nonpressure applications. Note that in the soft and hard tubing there are two types which have the same letter: types K and L. The wall thicknesses for these two types are the same.

METHODS OF MANUFACTURING PIPE

There are many ways of manufacturing steel pipe, such as seamless, lap welding, butt welding, and resistance welding (Fig. 4-1).

Seamless pipe: Just as the name implies, this is pipe without a seam. This is accomplished at the steel mill by piercing a solid billet with a piercing tool.

Lap-welded and butt-welded pipe: These types are manufactured by forming steel plate and then welding it together.

Resistance welding: This is a process of manufacturing pipe by passing an electric current through the material to obtain a welding point.

It should be noted that seamless pipe has greater strength than lap-welded pipe, which in turn has greater strength than butt-welded pipe. The electric-resistance pipe is not equal to seamless pipe.

All pipe shipped from a fabricator is shop-tested to ensure that it conforms to required specifications established by code and to customer specifications.

PIPE-SIZE FACTORS

Within the industry, pipe size is referred to by the term *nominal pipe size.* For all steel pipe ranging from ⅛" to 12", this means the approximate *inside diameter.* For all steel pipe 14" and above, it means

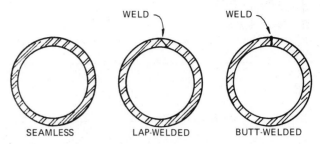

Fig. 4-1 Various methods of pipe manufacturing.

the exact *outside diameter.* Note that there is still the thickness of the pipe to be considered. This wall thickness is determined by the *weight* of the pipe. Pipe weight is subject to the design specifications for the particular service for which the pipe is intended. The design of the pipe must take into consideration such factors as the characteristics of the fluid or gas which is to be handled, such as its temperature, pressure, and erosion or corrosion factor, as well as economics, such as the initial cost against maintenance. The recognized codes for pressure piping within the industry must also be considered: American National Standards Institute (ANSI), American Society for Testing and Materials (ASTM), and American Society of Mechanical Engineers (ASME).

Since pipe sizes are standardized, tables such as Fig. 4-2 are used for dimensional requirements.

PIPE WEIGHT, OR MORE SPECIFICALLY PIPE SCHEDULES

The weight of steel pipe is referred to as the *schedule of pipe,* which may vary from schedule 10 to schedule 160. Any pipe manual such as Tube Turns, or the equivalent will give all the necessary dimensions once the schedule of pipe has been determined. Note in these dimensions that the wall thickness becomes greater as the schedule of pipe increases from schedule 10 to schedule 160 (see Fig. 4-2, p. 30).

When reference is made to standard-weight pipe (Std.), it is schedule 40 from ½" to 12" pipe size; extra-heavy pipe (X-hvy.) is schedule 80 from ½" to 10" pipe size; double-extra-heavy (XX-hvy.) varies from pipe size to pipe size.

The outside diameter of any size of pipe is the same for all weights. Variations in weight and wall thickness affect only the inside diameter. Pipe 12" and smaller is known by its nominal inside diameter; pipe 14" and larger is known by its outside diameter. Boiler tubes and tubing are known by their outside diameters. When calling out pipe 12" and smaller, designate the weight desired: standard, extra heavy, or double extra heavy. When calling out pipe having an outside diameter of 14" or larger, designate the wall thickness desired.

29

Dimensions of Seamless and Welded Pipe

ANSI B36.10 API Std. 5L and 5LX
NOMINAL WALL THICKNESSES FOR ALL SCHEDULES
(WITH −12.50% VALUES)

Nom. Size	Nom. OD	Sch. 5 ID	Wall	Sch. 10 ID	Wall	Gas Dist. ID	Wall	Sch. 20 ID	Wall	Std. ID	Wall	Sch. 30 ID	Wall	Sch. 40 ID	Wall	XH ID	Wall
⅛	.405	......	....	......	....	......	....	......	....	.269	.060 / .068	......	....	STD.		.215	.083 / .095
¼	.540	......	....	......		......	....	......	....	.364	.077 / .088	......	....	STD.		.302	.104 / .119
⅜	.675	......	....	......	....	......	....	......	....	.493	.080 / .091	......	....	STD.		.423	.110 / .126
½	.840	......	....	......	....	......	....	......	....	.622	.096 / .109	......	....	STD.		.546	.129 / .147
¾	1.050	.920	.057 / .065	.844	.073 / .083	......	....	......	....	.824	.099 / .113	......	....	STD.		.742	.135 / .154
1	1.315	1.185	.057 / .065	1.097	.096 / .109	......	....	......	....	1.049	.117 / .133	......	....	STD.		.957	.157 / .179
1¼	1.660	1.530	.057 / .065	1.442	.096 / .109	......	....	......	....	1.380	.123 / .140	......	....	STD.		1.278	.167 / .191
1½	1.900	1.770	.057 / .065	1.682	.096 / .109	......	....	......	....	1.610	.127 / .145	......	....	STD.		1.500	.175 / .200
2	2.375	2.245	.057 / .065	2.157	.096 / .109	......	....	......	....	2.067	.135 / .154	......	....	STD.		1.939	.191 / .218
2½	2.875	2.709	.073 / .083	2.635	.105 / .120	......	....	......	....	2.469	.178 / .203	......	....	STD.		2.323	.242 / .276
3	3.500	3.334	.073 / .083	3.260	.105 / .120	......	....	......	....	3.068	.189 / .216	......	....	STD.		2.900	.263 / .300
3½	4.000	3.834	.073 / .083	3.760	.105 / .120	......	....	......	....	3.548	.198 / .226	......	....	STD.		3.364	.279 / .318
4	4.500	4.334	.073 / .083	4.260	.105 / .120	4.124	.164 / .188	......	....	4.026	.207 / .237	......	....	STD.		3.826	.295 / .337
5	5.563	5.345	.095 / .109	5.295	.118 / .134	......	....	......	....	5.047	.226 / .258	......	....	STD.		4.813	.328 / .375
6	6.625	6.407	.095 / .109	6.357	.118 / .134	6.187	.192 / .219	......	....	6.065	.245 / .280	......	....	STD.		5.761	.378 / .432
8	8.625	8.407	.095 / .109	8.329	.130 / .148	8.187	.192 / .219	8.125	.219 / .250	7.981	.282 / .322	8.071	.242 / .277	STD.		7.625	.438 / .500
10	10.750	10.482	.117 / .134	10.420	.145 / .165	10.312	.192 / .219	10.250	.219 / .250	10.020	.320 / .365	10.136	.269 / .307	STD.		9.750	.438 / .500
12	12.750	12.438	.137 / .156	12.390	.158 / .180	12.250	.219 / .250	12.250	.219 / .250	12.000	.328 / .375	12.090	.289 / .330	11.938	.355 / .406	11.750	.438 / .500
14	14.000	......	....	13.500	.219 / .250	13.500	.219 / .250	13.376	.273 / .312	13.250	.328 / .375	STD.		13.124	.383 / .438	13.000	.438 / .500
16	16.000	......	....	15.500	.219 / .250	15.500	.219 / .250	15.376	.273 / .312	15.250	.328 / .375	STD.		XH		15.000	.438 / .500
18	18.000	......	....	17.500	.219 / .250	17.500	.219 / .250	17.376	.273 / .312	17.250	.328 / .375	17.124	.438	16.876	.492 / .562	17.000	.438 / .500
20	20.000	......	....	19.500	.219 / .250	19.500	.219 / .250	STD.		19.250	.328 / .375	XH		18.812	.519 / .594	19.000	.438 / .500
22	22.000	......	....	21.500	.219 / .250	......	....	STD.		21.250	.328 / .375	XH		......	....	21.000	.438 / .500
24	24.000	......	....	23.500	.219 / .250	23.500	.219 / .250	STD.		23.250	.328 / .375	22.876	.492 / .562	22.624	.602 / .688	23.000	.438 / .500
26	26.000			25.376	.273 / .312	......	....	XH		25.250	.328 / .375	......	....	......	....	25.000	.438 / .500
28	28.000			27.376	.273 / .312	......	....	XH		27.250	.328 / .375	26.750	.547 / .625	......	....	27.000	.438 / .500
30	30.000	......	....	29.376	.273 / .312	......	....	XH		29.250	.328 / .375	28.750	.547 / .625	......	....	29.000	.438 / .500
36	36.000	......	....	35.376	.273 / .312	......	....	XH		35.250	.328 / .375	34.750	.547 / .625	34.500	.656 / .750	35.000	.438 / .500
42	42.000	......	....	......	....	......	....	......	....	41.250	.328 / .375	40.750	.547 / .625	40.500	.656 / .750	41.000	.438 / .500

Fig. 4-2 Pipe size and schedules.

Dimensions of Seamless and Welded Pipe (continued)

Upper Figure in Wall Thickness Column is Minimum Wall. Lower Figure is Nominal Wall.

Sch. 60 ID	Wall	Sch. 80 ID	Wall	Sch. 100 ID	Wall	Sch. 120 ID	Wall	Sch. 140 ID	Wall	Sch. 160 ID	Wall	XXH ID	Wall	API ID	Wall	Nom. Size
......	....	XH		......	....	......	....	......	....	......	....	......	....	......	....	1/8
......	....	XH		......	....	......	....	......	....	......	....	......	....	......	....	1/4
......	....	XH		......	....	......	....	......	....	......	....	......	....	......	....	3/8
......	....	XH		......	....	......	....	......	....	.464	.165 / .188	.252	.257 / .294	......	....	1/2
......	....	XH		......	....	......	....	......	....	.612	.192 / .219	.434	.270 / .308	......	....	3/4
......	....	XH		......	....	......	....	......	....	.815	.219 / .250	.599	.313 / .358	......	....	1
......	....	XH		......	....	......	....	......	....	1.160	.219 / .250	.896	.334 / .382	......	....	1 1/4
......	....	XH		......	....	......	....	......	....	1.338	.246 / .281	1.100	.350 / .400	......	....	1 1/2
......	....	XH		......	....	......	....	......	....	1.687	.301 / .344	1.503	.382 / .436	......	....	2
......	....	XH		......	....	......	....	......	....	2.125	.328 / .375	1.771	.483 / .552	......	....	2 1/2
......	....	XH		......	....	......	....	......	....	2.624	.383 / .438	2.300	.525 / .600	3.250	.109 / .125	3
......	......	XH		......	....	......	....	......	....	......	....	2.728*	.557* / .636*	......	....	3 1/2
......	....	XH		......	....	3.624	.383 / .438	......	....	3.438	.465 / .531	3.152	.590 / .674	4.250	.109 / .125	4
......	....	XH		......	....	4.563	.438 / .500	......	....	4.313	.547 / .625	4.063	.656 / .750	......	....	5
......	....	XH		......	....	5.501	.492 / .562	......	....	5.187	.629 / .719	4.897	.756 / .864	6.313**	.137 / .156	6
7.813	.355 / .406	XH		7.437	.520 / .594	7.187	.629 / .719	7.001	.710 / .812	6.813	.793 / .906	6.875	.766 / .875	8.281**	.151 / .172	8
XH		9.562	.520 / .594	9.312	.629 / .719	9.062	.739 / .844	8.750	.875 / 1.000	8.500	.984 / 1.125	SCH 140		10.374	.165 / .188	10
11.626	.492 / .562	11.374	.602 / .688	11.062	.739 / .844	10.750	.875 / 1.000	10.500	.984 / 1.125	10.126	1.148 / 1.312	SCH 120		12.344	.178 / .203	12
12.812	.520 / .594	12.500	.656 / .750	12.124	.821 / .938	11.812	.957 / 1.094	11.500	1.094 / 1.250	11.188	1.231 / 1.406	......	....	......	....	14
14.688	.574 / .656	14.312	.739 / .844	13.938	.902 / 1.031	13.562	1.067 / 1.219	13.124	1.258 / 1.438	12.812	1.395 / 1.594	......	....	......	....	16
16.500	.656 / .750	16.124	.821 / .938	15.688	1.012 / 1.156	15.250	1.203 / 1.375	14.876	1.367 / 1.562	14.438	1.558 / 1.781	......	....	......	....	18
18.376	.711 / .812	17.938	.902 / 1.031	17.438	1.121 / 1.281	17.000	1.313 / 1.500	16.500	1.531 / 1.750	16.062	1.723 / 1.969	......	....	......	....	20
20.250	.766 / .875	19.750	.984 / 1.125	19.250	1.203 / 1.375	18.750	1.422 / 1.625	18.250	1.641 / 1.875	17.750	1.859 / 2.125	......	....	......	....	22
22.062	.848 / .969	21.562	1.067 / 1.219	20.938	1.340 / 1.531	20.376	1.586 / 1.812	19.876	1.804 / 2.062	19.312	2.051 / 2.344	......	....	......	....	24
......	....	......	....	......	....	......	....	......	....	......	....	......	....	......	....	26
......	....	......	....	......	....	......	....	......	....	......	....	......	....	......	....	28
......	....	......	....	......	....	......	....	......	....	......	....	......	....	......	....	30
......	....	......	....	......	....	......	....	......	....	......	....	......	....	......	....	36
......	....	......	....	......	....	......	....	......	....	......	....	......	....	......	....	42

*Not in ANSI B36.10   **Not in API

The oil companies of the United States have decided to call 1¼", 3½", and 5" pipe *nonstandard sizes*. For this reason these sizes are seldom if ever used except for special conditions, such as when required for connection to a piece of equipment which has that particular size connection.

END PREPARATION OF STEEL PIPE

Depending on the application, pipe can be purchased with either beveled, threaded, or plain ends (see Fig. 4-3).

Beveled: A pipe end is beveled when it is necessary to weld it to a fitting or another length of pipe (see Fig. 4-4).

Threaded: Pipe is threaded when the piping system is designed for threaded (screwed) connections (see Fig. 4-5).

Plain-End: Pipe is plain-end when the piping system is designed for socket-weld connections (see Fig. 4-6).

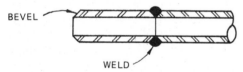

PLAIN END   BEVELED END   SCREWED OR THREADED END

Fig. 4-3   End preparation of pipe. Ends can be plain, beveled, or threaded.

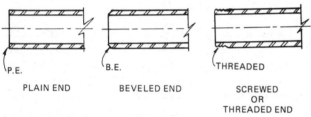

Fig. 4-4   Welded pipe fittings.

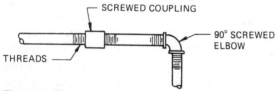

Fig. 4-5   Threaded pipe fittings.

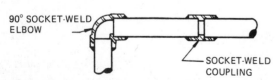

Fig. 4-6   Socket weld pipe fittings.

PIPE-BENDING STANDARDS AND APPLICATION

In certain situations it is more feasible to bend pipe than to use fittings such as elbows or returns. If proper methods of pipe bending are used, much time and money are saved with this method of fabrication. (See Fig. 4-7 for basic types of bends.)

Pipe bends are made by filling the pipe with sand, heating it carefully until the entire sand core produces a uniform temperature, then clamping the pipe to forms which have been accurately laid out on the bending floor. Power winches are applied to draw the pipe around the forms, and the point of power application is repeatedly shifted to prevent the formation of wrinkles on the inside wall (see Fig. 4-8, p. 33).

All bends are completed in the pipe before beveling, threading, or flanging, to assure accuracy. A system of repeated checking assures that all dimensions and angles of the completed pipe bend are accurate and that the ends or flanges are square with the axis of the pipe. Pipe bends are full-area and free from wrinkles, buckles, and thin spots; they are cleaned of all scale, chips, and oil. They are easy to erect, fitting into place exactly (see Fig. 4-9).

In order to conform to a required layout, steel pipe can be bent without losing its physical properties. The minimum bending radius most commonly used is *five times* the nominal diameter of the pipe.

CAST-IRON PIPE

Cast-iron pipe has characteristics quite different from steel pipe. There are two types of cast-iron pipe: *hub-and-spigot* (or *soil*) pipe, and *bell-and*

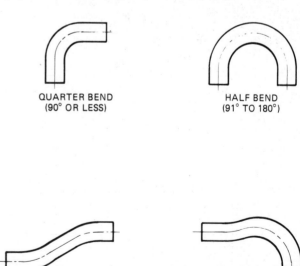

QUARTER BEND (90° OR LESS)   HALF BEND (91° TO 180°)

OFFSET BEND   SINGLE-OFFSET QUARTER BEND

Fig. 4-7   Basic types of bends.

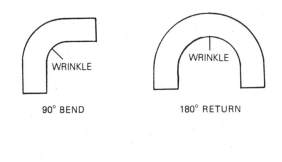

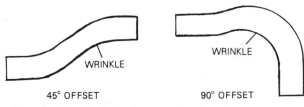

Fig. 4-8 Wrinkle formations on pipe bends.

Fig. 4-9 Pipe bending processes.

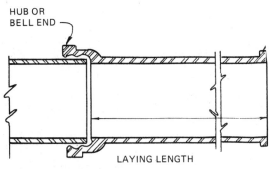

Fig. 4-10 Hub or bell and spigot pipes.

spigot (or *AWWA*—for American Water Works Association) pipe. The hub- and bell-and-spigot ends of bell types are similar. The main differences are in the wall thickness and laying length (see Fig. 4-10).

Hub-and-spigot or soil pipe is generally used for light service as in buildings, or where pressures are low as in low-pressure water systems and drainage systems.

Hub-and-spigot piping is never used where it may be subject to impact loads, such as under roadways or trucking areas. It is good practice never to place underground piping under foundations because it is difficult to make this location accessible for maintenance. The laying length for bell-and-spigot pipe is standardized at 16' 0". Cast-iron pipe will not take bending; as a result, all changes in direction are made with fittings.

These different types of cast-iron pipe also have their own fittings, known as *hub-and-spigot* or *soil fittings* and *bell-and-spigot* or *AWWA fittings*. As is the case with all pipe and fittings, manufacturers' catalogs furnish all the necessary types and dimensions required for a proper and economical layout.

METHODS OF DRAWING PIPE

Pipe can be drawn one of two ways, either single line or double line. *Single-line piping* is a schematic approach, with fittings as well as valves represented with symbols (see Fig. 4-11, p. 34). This is a faster method than double line and leaves the drawing less cluttered and congested. However, in order to be able to interpret this type of drawing, one must be familiar with symbols and pipe orientation.

Double-line piping is used primarily for larger pipe, with diameters of 12" and larger, and for situations where a particular portion of the pipe system needs to be shown in more detail (Fig. 4-12, p. 34). However, valves are still shown schematically because of the complexity of many valves.

Whether pipe is drawn single line or double line, dimensions are usually given to centerline of pipe. An exception would be if a pipe was resting on a pipe rack or support, so that it was necessary to give dimensions to bottom of pipe or top of steel.

From a pipe-size chart it is evident that the nominal pipe size and the actual pipe outside diameter (O.D.) are usually not the same. A pipe drafter must constantly refer to a size chart to help in the dimensioning process (see Figs. 4-13a and b, p. 35).

LINE DESIGNATION

Each pipe run in a process pipeline has its own line designation, and it is the only pipe with that

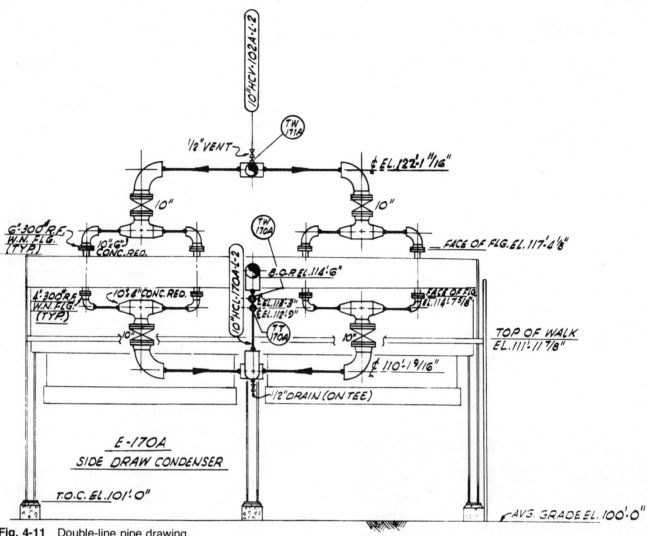

Fig. 4-11 Double-line pipe drawing.

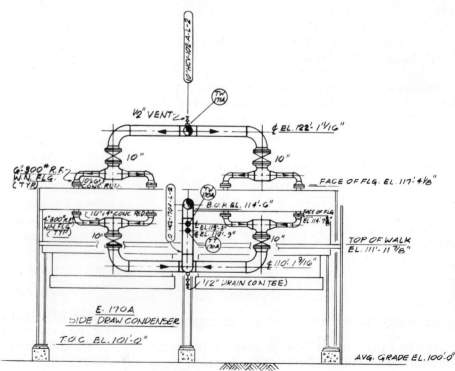

Fig. 4-12 Single-line pipe drawing.

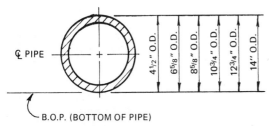

Fig. 4-13a Pipe diameters.

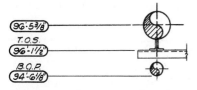

Fig. 4-13b Diameters affect dimensions.

particular identification. Very briefly, a line is designated as follows:

8″–HA–101–M

8″	indicated line size
HA	type of service
101	the area number
M	specifications

Many companies have their own method of line designation. Basically the same information must be given; however, it may not be in this sequence. A line number can change when one of the following things occurs: (1) size changes, (2) specifications change, (3) a branch leaves an original line, or (4) an additional line number would simplify the engineering and drawing (see Fig. 4-14).

In addition to major line designation for pipe runs, lines are further divided into pipe-fabrication spools. The spools of a particular line carry the major line identification as well as a dash number. Dash numbers begin upstream with the line designation plus dash 1, and progress downstream with dash numbers getting larger the further downstream they go (see Fig. 4-15).

The line number is basically 6″–HA–101–M. It changes to an 8″, then reduces again to a 6″, but basically in this situation the line number does not change.

The spools are designated as follows (starting upstream):

6″–HA–101–M–1
8″–HA–101–M–2
8″–HA–101–M–3
8″–HA–101–M–4

Spool sequence

All the situations for determining the line numbers would take a great deal of space, but if the basic

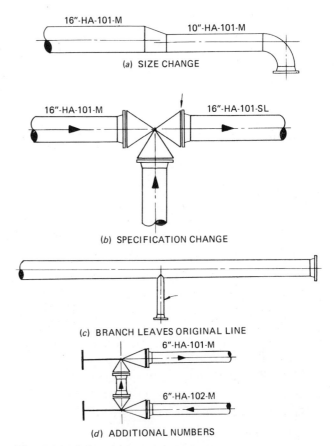

Fig. 4-14 A line designation change.

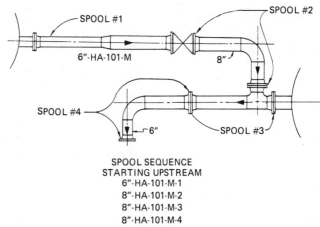

SPOOL SEQUENCE
STARTING UPSTREAM
6″-HA-101-M-1
8″-HA-101-M-2
8″-HA-101-M-3
8″-HA-101-M-4

Fig. 4-15 Dash numbering system.

explanations are followed, determining line numbers should come rather easily.

CONCLUSION

As can be determined from the foregoing discussion, pipe is a very important part of the total piping system and must be treated with a great degree of care.

EXERCISES

4-1. Define the following terms:

a. Pipe
b. Carbon steel
c. Alloy steel
d. Stainless steel
e. Cast iron
f. Tubing
g. Steam tracing
h. Seamless pipe
i. Lap weld
j. Butt weld
k. Resistance-welding process
l. Plain-end pipe
m. Beveled end
n. Threaded end
o. Pipe bending
p. Expansion U bend
q. Hub pipe
r. Spigot pipe
s. Schedule of pipe
t. Nominal pipe size
u. Fabrication
v. Field weld

4-2. Match the terms and abbreviations:

1. B.O.P.
2. T.O.S.
3. C_L
4. C.I. pipe
5. smls. pipe
6. specs
7. N.P.S.
8. sch.
9. wt.
10. P.E.
11. B.E.
12. thrd.
13. E.R.W.
14. stl.
15. I.D.
16. O.D.
17. T.E.
18. AWWA
19. F.W.

a. Centerline
b. Seamless pipe
c. Specifications
d. Cast-iron pipe
e. Top of steel
f. Weight
g. Bevel end
h. Electric-resistance weld
i. Steel
j. Threaded
k. Plain end
l. Nominal pipe size
m. Threaded end
n. Inside diameter
o. Field weld
p. American Water Works Association
q. Outside diameter
r. Schedule
s. Bottom of pipe

4-3. Write a three-page report on pipe-manufacturing methods. Use reference materials other than text material.

4-4. Write a two-page report on ancient pipe systems and their gradual development to present-day systems.

4-5. Make a list of as many situations as you can find where pipe is used.

4-6. Write a report on the methods of pipe bending. Use pictures, diagrams, etc., to explain the process.

DRAWING PROBLEMS

4-7. Make a one-line pipe drawing (see Fig. 4-11).

4-8. Make a two-line pipe drawing (see Fig. 4-12).

4-9. Instructor: Make drawing assignment from a pipe drawing in Appendix.

Five

Flanges and Flange Selection

Flanges vary widely in size, function, and type, but each basically enables a run of pipe to be taken apart by unbolting a series of bolts around the flange's outside diameter. After the flange is taken apart, valves, pipe, or other fittings can be repaired or replaced. Flanges, much the same as pipe, have definite standardized sizes and methods of application. Standardization of types, sizes, and application will be emphasized in the chapter in an effort to further prepare the pipe drafter to function effectively.

IMPORTANCE OF USING FLANGES

Flanges are used primarily as a method of connecting valves and fittings to pipe. Flanges enable valves to be removed periodically for repair or replacement by simple removal of several bolts. This eliminates the costly process of cutting into a system and having to thread or weld it back. Flanges have many other uses, which will be discussed in this chapter.

TYPES OF FORGED-STEEL FLANGES

There are six major types of forged-steel flanges, each of which has an advantage for a particular application. Study and become familiar with each.

Screwed Flange, Seal-welded (see Fig. 5-1):
A forged-steel screwed flange is used in this joint. The pipe and the flange are accurately threaded; the flange is made up tight on the pipe, seal-welded, and then refaced. The joint is sealed by fillet welding the back of the flange to the pipe, thus assuring no leakage through the threads.

The refacing assures perfect alignment of the flange faces, and that the end of the pipe is flush with the face of the flange for a proper gasket-bearing surface to the inside of the pipe. The threads retain the function of holding the flange securely on the pipe; hence, there is no shearing action.

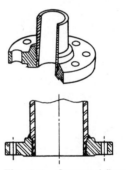

Fig. 5-1 Screwed flange.

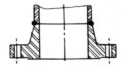

Fig. 5-2 Welding neck flange.

Weld-neck Flange (see Fig. 5-2): Weld-neck flanges are forged steel. They are machined with a beveled end and bored to match the inside diameter of the pipe to which they are applied. A butt weld is used to attach the weld-neck flange to the pipe, which is also machine-beveled.

Slip-on Welding Flange: Forged-steel slip-on welding flanges are bored for a snug fit on the pipe. When these flanges are applied to fabricated piping, they are welded at the front and back through two methods.

Type 1 is standard procedure for welded flanged joints on carbon-steel and genuine wrought-iron piping with 150- and 300-lb forged-steel slip-on welding flanges. Regular flanges are used with the end of the pipe set back from the face of the flange and the flange welded to the pipe both front and back (see Fig. 5-3).

Type 2 is standard procedure for welded flanged joints on carbon-steel and genuine wrought-iron piping with 400-lb or higher pressure class forged-carbon-steel flanges, and for all iron-pipe-size alloy piping where alloy flanges of all pressure classes are used. The pipe is flush with the flange face; this is accomplished by refacing after both the front and back of the flange are welded to the pipe (see Fig. 5-4).

Code Limitation: When piping must comply with either the pressure piping code or the ASME unfired pressure vessel code, the use of the slip-on welding flange joint is limited to a primary-service pressure rating of 300 lb/in².

Slip-on flanges are not commonly used in refineries for many reasons, one being that their strength is only about two-thirds that of a welding-neck flange; the ANSI and the ASME limit slip-on flanges to 150- and 300-lb classes. Although the initial cost of slip-on flanges is lower than that of a welding neck, when installed, slip-on flanges may cost only a little less than a welding neck. There are, however, some advantages to the use of the slip-on flange. When makeup conditions are tight, the slip-on flange uses less room than a welding-neck flange, and when alignment is difficult, the fieldwork is simplified. The drilling and facing of a slip-on flange are the same as those of a weld-neck flange in the same class.

Socket-weld Flange (see Fig. 5-5):   The socket-weld flange has a recess in the hub to receive the end of the pipe, and it is attached by means of a single heavy fillet weld. This type of flange is generally limited to smaller pipe sizes and lower pressure ratings. The initial cost is 10 percent greater than for a slip-on. However, where a customer requires that screwed connections be seal-welded, the use of socket-weld flanges and fittings is recommended for economy.

Fig. 5-3   Type-1 slip-on flange.

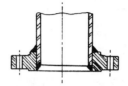

Fig. 5-4   Type-2 slip-on flange.

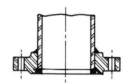

Fig. 5-5   Socket weld flange.

Blind Flange (see Fig. 5-6):   A blind flange is used for blanking off or blinding off ends of pipe. (See p. 38.) It is sometimes called a *blank flange.*

End-and-lap-joint Flange (see Fig. 5-7):   The lap-stub-end-and-lap flange can be applied to fabricated piping. Both the stub end and the pipe are machine-beveled. A butt weld is used to complete the joint.

SEVEN PRESSURE RATINGS FOR FORGED-STEEL FLANGES

The seven pressure ratings are as follows (see Fig. 5-8, p. 40): 150 lb, 300 lb, 400 lb, 600 lb, 900 lb, 1,500 lb, 2,500 lb. Each of the pressure ratings has specific dimensions that must be considered in fitting up flanges to various other fittings and valves.

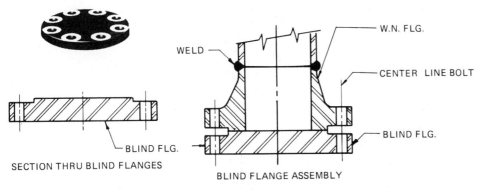

Fig. 5-6 Blind flange.

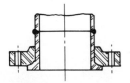

Fig. 5-7 Lap flange.

USE FLANGE CHART TO DETERMINE FLANGE DIMENSIONS

The weld-neck-flange bore chart (see Fig. 5-9, p. 40) is used as a quick reference for determining the bore and outside diameter of a flange. Simply find the nominal pipe size at left and the schedule at top. Come down and across and read bore in inches and tenths of an inch (see Fig. 5-10, p. 41).

SELECT NUMBER AND BOLT SIZE FROM CHART

Pictured in Fig. 5-11 is a portion of the chart shown in Fig. 5-8. For simplicity of instruction only, the 150- and 300-lb class was selected.

The number of bolt holes in a flange varies with the size and class. For example, in the 150-lb class a 3" flange has four bolt holes, and a 6" flange has eight bolt holes. In the 300-lb class the 3" flange has eight bolt holes and the 6" flange has twelve bolt holes (see Fig. 5-11, p. 41).

The diameter of the bolt hole also varies with class in order that a large-diameter bolt may be used for additional strength in the heavier classes. In every class, the diameter of the bolt is slightly smaller than the diameter of the bolt hole.

It is important to remember that when flanges are assembled to pipe, the bolts always straddle the nor-mal perpendicular or horizontal centerline (see Fig. 5-12, p. 41). The purpose of this is to prevent holes from mismatching at the erection site. If such mismatching were to occur on a job, it could be a very costly mistake.

The sectional view pictured in Fig. 5-13 (p. 41) may help clarify the fit-up procedure for flanges and how matching bolt circles must be considered. The flanges pictured in Fig. 5-13 are weld-neck, but the principle is the same for all other types of flanges.

Flanges are bolted together normally with either machine bolts or stud bolts (see Fig. 5-14, p. 42).

Clearly, when the class of flange increases from 150 to 400 lbs, the thickness increases, causing the machine bolt and stud bolt length dimension (represented by a and b) to increase. This condition is very often found when mating a flange to a piece of equipment such as a pump, compressor, condensor, filter or relief valve.

TYPES OF FLANGE FACINGS

There is a variety of facings and finishes for flanged joints. The word *facing* refers to the type of face that is machined on flanges or lap joints. It should not be confused with the word *finish*, which refers to the kind of finish that is applied to the actual contact surface of the face. For example, the facing on a flange machined with a ¼"-high projection is commonly known as a ¼"-*high raised face*, but the finish on the contact surface of the raised face projection can be either *serrated* or *smooth*.

Ring-type-joint Flange Facing (R.T.J.): When an R.T.J. facing is applied to a 150- or 300-lb valve or fitting, the center-to-raised-face dimension is equal to the regular center-to-face dimension (which includes the ¹⁄₁₆" raised face), plus the depth of the groove on sizes 2" and larger, or plus the depth of the groove minus ¹⁄₃₂" on smaller sizes (see Fig. 5-15, p. 42). (*Center* to raised face means center valve or flanged fitting. In reference to a weld-neck flange, use dimension Y; see Fig. 5-16, p. 42).

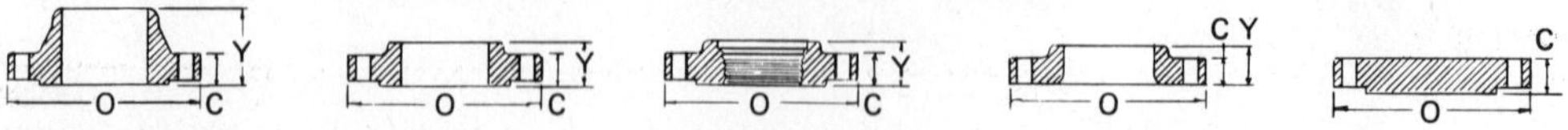

WELD NECK FLANGE | SLIP-ON FLANGE | THREADED FLANGE | LAP JOINT FLANGE | BLIND FLANGE

150 LB. FLANGES

Nom. Pipe Size	O	C②	Y② Weld Neck	Y② Slip On Thrd.	Y② Lap Joint	Bolt Circle	No. and Size of Holes
1/2	3 1/2	7/16	1 7/8	5/8	5/8	2 3/8	4-5/8
3/4	3 7/8	1/2	2 1/16	5/8	5/8	2 3/4	4-5/8
1	4 1/4	9/16	2 3/16	11/16	11/16	3 1/8	4-5/8
1 1/4	4 5/8	5/8	2 1/4	13/16	13/16	3 1/2	4-5/8
1 1/2	5	11/16	2 7/16	7/8	7/8	3 7/8	4-5/8
2	6	3/4	2 1/2	1	1	4 3/4	4-3/4
2 1/2	7	7/8	2 3/4	1 1/8	1 1/8	5 1/2	4-3/4
3	7 1/2	15/16	2 3/4	1 3/16	1 3/16	6	4-3/4
3 1/2	8 1/2	15/16	2 13/16	1 1/4	1 1/4	7	8-3/4
4	9	15/16	3	1 5/16	1 5/16	7 1/2	8-3/4
5	10	15/16	3 1/2	1 7/16	1 7/16	8 1/2	8-7/8
6	11	1	3 1/2	1 9/16	1 9/16	9 1/2	8-7/8
8	13 1/2	1 1/8	4	1 3/4	1 3/4	11 3/4	8-7/8
10	16	1 3/16	4	1 15/16	1 15/16	14 1/4	12-1
12	19	1 1/4	4 1/2	2 3/16	2 3/16	17	12-1
14	21	1 3/8	5	2 1/4	3 1/8	18 3/4	12-1 1/8
16	23 1/2	1 7/16	5	2 1/2	3 7/16	21 1/4	16-1 1/8
18	25	1 9/16	5 1/2	2 11/16	3 13/16	22 3/4	16-1 1/4
20	27 1/2	1 11/16	5 11/16	2 7/8	4 1/16	25	20-1 1/4
24	32	1 7/8	6	3 1/4	4 3/8	29 1/2	20-1 3/8
30	38 3/4	2 1/8	5 1/8	3 1/2	..	36	28-1 3/8
36	46	2 3/8	5 3/8	3 3/4	..	42 3/4	32-1 5/8

300 LB. FLANGES

Nom. Pipe Size	O	C②	Y② Weld Neck	Y② Slip on Thrd.	Y② Lap Joint	Bolt Circle	No. and Size of Holes
1/2	3 3/4	9/16	2 1/16	7/8	7/8	2 5/8	4-5/8
3/4	4 5/8	5/8	2 1/4	1	1	3 1/4	4-3/4
1	4 7/8	11/16	2 7/16	1 1/16	1 1/16	3 1/2	4-3/4
1 1/4	5 1/4	3/4	2 9/16	1 1/16	1 1/16	3 7/8	4-3/4
1 1/2	6 1/8	13/16	2 11/16	1 3/16	1 3/16	4 1/2	4-7/8
2	6 1/2	7/8	2 3/4	1 5/16	1 5/16	5	8-3/4
2 1/2	7 1/2	1	3	1 1/2	1 1/2	5 7/8	8-7/8
3	8 1/4	1 1/8	3 1/8	1 11/16	1 11/16	6 5/8	8-7/8
3 1/2	9	1 3/16	3 3/16	1 3/4	1 3/4	7 1/4	8-7/8
4	10	1 1/4	3 3/8	1 7/8	1 7/8	7 7/8	8-7/8
5	11	1 3/8	3 7/8	2	2	9 1/4	8-7/8
6	12 1/2	1 7/16	3 7/8	2 1/16	2 1/16	10 5/8	12-7/8
8	15	1 5/8	4 3/8	2 7/16	2 7/16	13	12-1
10	17 1/2	1 7/8	4 5/8	2 5/8	3 3/4	15 1/4	16-1 1/8
12	20 1/2	2	5 1/8	2 7/8	4	17 3/4	16-1 1/4
14	23	2 1/8	5 5/8	3	4 3/8	20 1/4	20-1 1/4
16	25 1/2	2 1/4	5 3/4	3 1/4	4 3/4	22 1/2	20-1 3/8
18	28	2 3/8	6 1/4	3 1/2	5 1/8	24 3/4	24-1 3/8
20	30 1/2	2 1/2	6 3/8	3 3/4	5 1/2	27	24-1 3/8
24	36	2 3/4	6 5/8	4 3/16	6	32	24-1 5/8
30	43	3 3/8	8 1/4	8 1/4	..	39 1/4	28-1 7/8
36	50	4 1/8	9 1/2	9 1/2	..	46	32-2 1/8

400 LB. FLANGES

Nom. Pipe Size	O	C②	Y② Weld Neck	Y② Slip on Thrd.	Y② Lap Joint	Bolt Circle	No. and Size of Holes
1/2	3 3/4	9/16	2 1/16	7/8	7/8	2 5/8	4-5/8
3/4	4 5/8	5/8	2 1/4	1	1	3 1/4	4-3/4
1	4 7/8	11/16	2 7/16	1 1/16	1 1/16	3 1/2	4-3/4
1 1/4	5 1/4	13/16	2 5/8	1 1/8	1 1/8	3 7/8	4-3/4
1 1/2	6 1/8	7/8	2 3/4	1 1/4	1 1/4	4 1/2	4-7/8
2	6 1/2	1	2 7/8	1 7/16	1 7/16	5	8-3/4
2 1/2	7 1/2	1 1/8	3 1/8	1 5/8	1 5/8	5 7/8	8-7/8
3	8 1/4	1 1/4	3 1/4	1 13/16	1 13/16	6 5/8	8-7/8
3 1/2	9	1 3/8	3 3/8	1 15/16	1 15/16	7 1/4	8-1
4	10	1 3/8	3 1/2	2	2	7 7/8	8-1
5	11	1 1/2	4	2 1/8	2 1/8	9 1/4	8-1
6	12 1/2	1 5/8	4 1/16	2 1/4	2 1/4	10 5/8	12-1
8	15	1 7/8	4 5/8	2 11/16	2 11/16	13	12-1 1/8
10	17 1/2	2 1/8	4 7/8	2 7/8	4	15 1/4	16-1 1/4
12	20 1/2	2 1/4	5 3/8	3 1/8	4 1/4	17 3/4	16-1 3/8
14	23	2 3/8	5 7/8	3 5/16	4 5/8	20 1/4	20-1 3/8
16	25 1/2	2 1/2	6	3 11/16	5	22 1/2	20-1 1/2
18	28	2 5/8	6 1/2	3 7/8	5 3/8	24 3/4	24-1 1/2
20	30 1/2	2 3/4	6 5/8	4	5 3/4	27	24-1 5/8
24	36	3	6 7/8	4 1/2	6 1/4	32	24-1 7/8
30	43	4	8 5/8	8 5/8	..	39 1/4	28-2 1/8
36	50	4 1/2	9 7/8	9 7/8	..	46	32-2 1/8

600 LB. FLANGES

Nom. Pipe Size	O	C②	Y② Weld Neck	Y② Slip on Thrd.	Y② Lap Joint	Bolt Circle	No. and Size of Holes
1/2	3 3/4	9/16	2 1/16	7/8	7/8	2 5/8	4-5/8
3/4	4 5/8	5/8	2 1/4	1	1	3 1/4	4-3/4
1	4 7/8	11/16	2 7/16	1 1/16	1 1/16	3 1/2	4-3/4
1 1/4	5 1/4	13/16	2 5/8	1 1/8	1 1/8	3 7/8	4-3/4
1 1/2	6 1/8	7/8	2 3/4	1 1/4	1 1/4	4 1/2	4-7/8
2	6 1/2	1	2 7/8	1 7/16	1 7/16	5	8-3/4
2 1/2	7 1/2	1 1/8	3 1/8	1 5/8	1 5/8	5 7/8	8-7/8
3	8 1/4	1 1/4	3 1/4	1 13/16	1 13/16	6 5/8	8-7/8
3 1/2	9	1 3/8	3 3/8	1 15/16	1 15/16	7 1/4	8-1
4	10 3/4	1 1/2	4	2 1/8	2 1/8	8 1/2	8-1
5	13	1 3/4	4 1/2	2 3/8	2 3/8	10 1/2	8-1 1/8
6	14	1 7/8	4 5/8	2 5/8	2 5/8	11 1/2	12-1 1/8
8	16 1/2	2 3/16	5 1/4	3	3	13 3/4	12-1 1/4
10	20	2 1/2	6	3 3/8	4 3/8	17	16-1 3/8
12	22	2 5/8	6 1/8	3 5/8	4 5/8	19 1/4	20-1 3/8
14	23 3/4	2 3/4	6 1/2	3 11/16	5	20 3/4	20-1 1/2
16	27	3	7	4 3/16	5 1/2	23 3/4	20-1 5/8
18	29 1/4	3 1/4	7 1/4	4 5/8	6	25 3/4	20-1 3/4
20	32	3 1/2	7 1/2	5	6 1/2	28 1/2	24-1 3/4
24	37	4	8	5 1/2	7 1/4	33	24-2
30	44 1/2	4 1/2	9 3/4	9 3/4	..	40 1/4	28-2 1/8
36	51 3/4	4 7/8	11 1/8	11 1/8	..	47	28-2 5/8

900 LB. FLANGES

Nom. Pipe Size	O	C②	Y② Weld Neck	Y② Slip on Thrd.	Y② Lap Joint	Bolt Circle	No. and Size of Holes
1/2	4 3/4	7/8	2 3/8	1 1/4	1 1/4	3 1/4	4-7/8
3/4	5 1/8	1	2 3/4	1 3/8	1 3/8	3 1/2	4-7/8
1	5 7/8	1 1/8	2 7/8	1 5/8	1 5/8	4	4-1
1 1/4	6 1/4	1 1/8	2 7/8	1 5/8	1 5/8	4 3/8	4-1
1 1/2	7	1 1/4	3 1/4	1 3/4	1 3/4	4 7/8	4-1 1/8
2	8 1/2	1 1/2	4	2 1/4	2 1/4	6 1/2	8-1
2 1/2	9 5/8	1 5/8	4 1/8	2 1/2	2 1/2	7 1/2	8-1 1/8
3	9 1/2	1 1/2	4	2 1/8	2 1/8	7 1/2	8-1
3 1/2	..	..	..	..	..	..	..
4	11 1/2	1 3/4	4 1/2	2 3/4	2 3/4	9 1/4	8-1 1/4
5	13 3/4	2	5	3 1/8	3 1/8	11	8-1 3/8
6	15	2 3/16	5 1/2	3 3/8	3 3/8	12 1/2	12-1 1/4
8	18 1/2	2 1/2	6 3/8	4	4 1/2	15 1/2	12-1 1/2
10	21 1/2	2 3/4	7 1/4	4 1/4	5	18 1/2	16-1 1/2
12	24	3 1/8	7 7/8	4 5/8	5 5/8	21	20-1 1/2
14	25 1/4	3 3/8	8 3/8	5 1/8	6 1/8	22	20-1 5/8
16	27 3/4	3 1/2	8 1/2	5 1/4	6 1/2	24 1/4	20-1 3/4
18	31	4	9	6	7 1/2	27	20-2
20	33 3/4	4 1/4	9 3/4	6 1/4	8 1/4	29 1/2	20-2 1/8
24	41	5 1/2	11 1/2	8	10 1/2	35 1/2	20-2 5/8
30	48 1/2	5 7/8	12 1/4	12 1/4	..	42 3/4	20-3 1/8
36	57 1/2	6 3/4	14 1/4	14 1/4	..	50 3/4	20-3 5/8

1500 LB. FLANGES

Nom. Pipe Size	O	C②	Y② Weld Neck	Y② Slip on Thrd.	Y② Lap Joint	Bolt Circle	No. and Size of Holes
1/2	4 3/4	7/8	2 3/8	1 1/4	1 1/4	3 1/4	4-7/8
3/4	5 1/8	1 7/8	2 3/4	1 3/8	1 3/8	3 1/2	4-7/8
1	5 7/8	1 1/8	2 7/8	1 5/8	1 5/8	4	4-1
1 1/4	6 1/4	1 1/8	2 7/8	1 5/8	1 5/8	4 3/8	4-1
1 1/2	7	1 1/4	3 1/4	1 3/4	1 3/4	4 7/8	4-1 1/8
2	8 1/2	1 1/2	4	2 1/4	2 1/4	6 1/2	8-1
2 1/2	9 5/8	1 5/8	4 1/8	2 1/2	2 1/2	7 1/2	8-1 1/8
3	10 1/2	1 7/8	4 5/8	2 7/8	2 7/8	8	8-1 1/4
3 1/2	..	..	..	..	..	..	..
4	12 1/4	2 1/8	4 7/8	3 9/16	3 9/16	9 1/2	8-1 3/8
5	14 3/4	2 7/8	6 1/8	4 1/8	4 1/8	11 1/2	8-1 5/8
6	15 1/2	3 1/4	6 3/4	4 11/16	4 11/16	12 1/2	12-1 1/2
8	19	3 5/8	8 3/8	5 5/8	5 5/8	15 1/2	12-1 3/4
10	23	4 1/4	10	6 1/4	7	19	12-2
12	26 1/2	4 7/8	11 1/8	7 1/8	8 5/8	22 1/2	16-2 1/8
14	29 1/2	5 1/4	11 3/4	..	9 1/2	25	16-2 3/8
16	32 1/2	5 3/4	12 1/4	..	10 1/4	27 3/4	16-2 5/8
18	36	6 3/8	12 7/8	..	10 7/8	30 1/2	16-2 7/8
20	38 3/4	7	14	..	11 1/2	32 3/4	16-3 1/8
24	46	8	16	..	13	39	16-3 5/8
30	..	..	..	..	..	..	..
36	..	..	..	..	..	..	..

2500 LB. FLANGES

Nom. Pipe Size	O	C②	Y② Weld Neck	Y② Slip on Thrd.	Y② Lap Joint	Bolt Circle	No. and Size of Holes
1/2	5 1/4	1 3/16	2 7/8	1 9/16	1 9/16	3 1/2	4-7/8
3/4	5 1/2	1 1/4	3 1/8	1 11/16	1 11/16	3 3/4	4-7/8
1	6 1/4	1 3/8	3 1/2	1 7/8	1 7/8	4 1/4	4-1
1 1/4	7 1/4	1 1/2	3 3/4	2 1/16	2 1/16	5 1/8	4-1 1/8
1 1/2	8	1 3/4	4 3/8	2 3/8	2 3/8	5 3/4	4-1 1/4
2	9 1/4	2	5	2 3/4	2 3/4	6 3/4	8-1 1/8
2 1/2	10 1/2	2 1/4	5 5/8	3 1/8	3 1/8	7 3/4	8-1 1/4
3	12	2 5/8	6 5/8	3 5/8	3 5/8	9	8-1 3/8
4	14	3	7 1/2	4 1/4	4 1/4	10 3/4	8-1 5/8
5	16 1/2	3 5/8	9	5 1/8	5 1/8	12 3/4	8-1 7/8
6	19	4 1/4	10 3/4	6	6	14 1/2	8-2 1/8
8	21 3/4	5	12 1/2	7	7	17 1/4	12-2 1/8
10	26 1/2	6 1/2	16 1/2	9	9	21 1/4	12-2 5/8
12	30	7 1/4	18 1/4	10	10	24 3/8	12-2 7/8

WELDING NECK FLANGE BORES ①③

Nom. Pipe Size	Outside Diam.	Light Wall⑥	Sched. 20	Sched. 30	Std. Wall.	Sched. 40	Sched. 60	Extra Strong	Sched. 80	Sched. 100	Sched. 120	Sched. 140	Sched. 160	Double Extra Strong
1/2	0.840	.674	...	...	0.622	0.622	...	0.546	0.546	...	...	...	0.464	0.252
3/4	1.050	.884	...	...	0.824	0.824	...	0.742	0.742	...	...	...	0.612	0.434
1	1.315	1.097	...	...	1.049	1.049	...	0.957	0.957	...	...	...	0.815	0.599
1 1/4	1.660	1.442	...	...	1.380	1.380	...	1.278	1.278	...	...	...	1.160	0.896
1 1/2	1.900	1.682	...	...	1.610	1.610	...	1.500	1.500	...	...	...	1.338	1.100
2	2.375	2.157	...	...	2.067	2.067	...	1.939	1.939	...	...	...	1.687	1.503
2 1/2	2.875	2.635	...	...	2.469	2.469	...	2.323	2.323	...	...	...	2.125	1.771
3	3.500	3.260	...	...	3.068	3.068	...	2.900	2.900	...	...	...	2.624	2.300
3 1/2	4.000	3.760	...	...	3.548	3.548	...	3.364	3.364	...	...	...	...	2.728
4	4.500	4.260	...	...	4.026	4.026	...	3.826	3.826	...	3.624	...	3.438	3.152
5	5.563	5.295	...	...	5.047	5.047	...	4.813	4.813	...	4.563	...	4.313	4.063
6	6.625	6.357	...	...	6.065	6.065	...	5.761	5.761	...	5.501	...	5.187	4.897
8	8.625	8.329	8.125	8.071	7.981	7.981	7.813	7.625	7.625	7.437	7.187	7.001	6.813	6.875
10	10.750	10.420	10.250	10.136	10.020	10.020	9.750	9.750	9.562	9.312	9.062	8.750	8.500	8.750
12	12.750	12.390	12.250	12.090	12.000	11.938	11.626	11.750	11.374	11.062	10.750	10.500	10.126	10.750
14	14.000	13.500	13.376	13.250	13.250	13.124	12.812	13.000	12.500	12.124	11.814	11.500	11.188	...
16	16.000	15.500	15.376	15.250	15.250	15.000	14.688	15.000	14.312	13.938	13.564	13.124	12.812	...
18	18.000	17.500	17.376	17.124	17.250	16.876	16.500	17.000	16.124	15.688	15.250	14.876	14.438	...
20	20.000	19.500	19.250	19.000	19.250	18.812	18.376	19.000	17.938	17.438	17.000	16.500	16.062	...
24	24.000	23.500	23.250	22.876	23.250	22.624	22.062	23.000	21.562	20.938	20.376	19.876	19.312	...

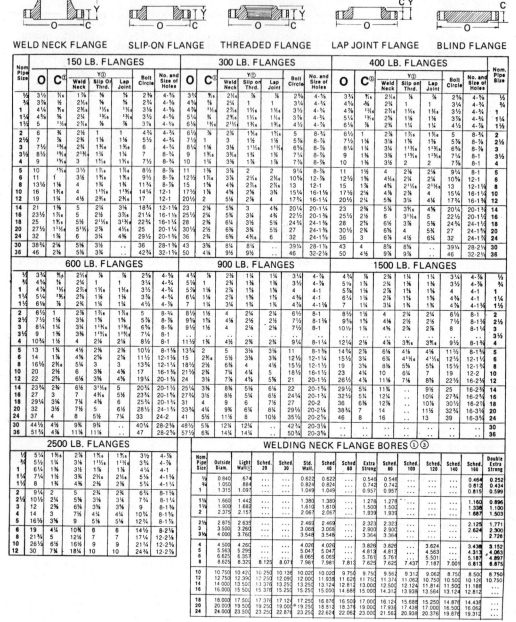

Fig. 5-8 Flange rating chart.

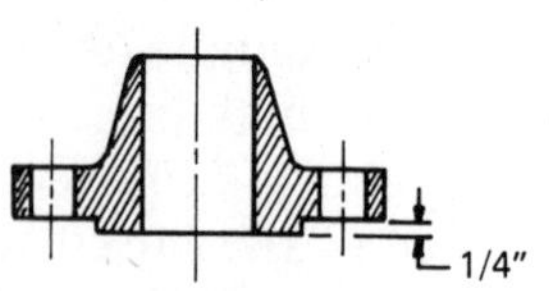

400# AND ABOVE W.N. FLG.

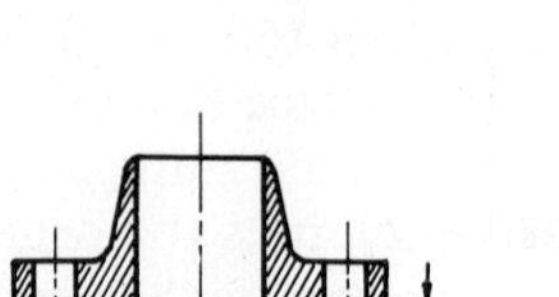

150# AND 300# W.N. FLG.

Fig. 5-9 Weld neck flange dimensions.

Tongue-and-groove or Male-and-female Facings: Tongue-and-groove and male-and-female facings have characteristics of their own: The groove or female facing is a groove in the raised face of a flange, and the tongue or male facing is a protrusion from the face of the flange. Figure 5-17 (p. 43) shows the extent of the facings. One characteristic to be noted is that the male or tongue facing increases the regular center-to-face dimension of the flange by 1/4". The female or groove facing does not alter the regular center-to-face dimension.

Special Flanges and Their Application: There are several types of special flanges. These are

Welding Neck Flange Bores [1] [3]

Nom. Pipe Size	Outside Diam.	Light Wall [6]	Sched. 20	Sched. 30	Std. Wall.	Sched. 40	Sched. 60	Extra Strong	Sched. 80	Sched. 100	Sched. 120	Sched. 140	Sched. 160	Double Extra Strong
½	0.840	.674	...	...	0.622	0.622	...	0.546	0.546	...	...	...	0.464	0.252
¾	1.050	.884	...	...	0.824	0.824	...	0.742	0.742	...	...	...	0.612	0.434
1	1.315	1.097	...	...	1.049	1.049	...	0.957	0.957	...	...	...	0.815	0.599
1¼	1.660	1.442	...	...	1.380	1.380	...	1.278	1.278	...	...	...	1.160	0.896
1½	1.900	1.682	...	...	1.610	1.610	...	1.500	1.500	...	...	...	1.338	1.100
2 2.375	2.157	...	...	2.067	2.067	...	1.939	1.939	...	...	...	1.687	1.503	
2½	2.875	2.635	...	...	2.469	2.469	...	2.323	2.323	...	...	...	2.125	1.771
3	3.500	3.260	...	...	3.068	3.068	...	2.900	2.900	...	...	...	2.624	2.300
3½	4.000	3.760	...	...	3.548	3.548	...	3.364	3.364	...	...	...	...	2.728
4	4.500	4.260	...	...	4.026	4.026	...	3.826	3.826	....	3.624	...	3.438	3.152
5	5.563	5.295	...	...	5.047	5.047	...	4.813	4.813	...	4.563	...	4.313	4.063
6	6.625	6.357	...	...	6.065	6.065	...	5.761	5.761	...	5.501	...	5.187	4.897
8	8.625	8.329	8.125	8.071	7.981	7.981	7.813	7.625	7.625	7.437	7.187	7.001	6.813	6.875
10	10.750	10.420	10.250	10.136	10.020	10.020	9.750	9.750	9.562	9.312	9.062	8.750	8.500	8.750
12	12.750	12.390	12.250	12.090	12.000	11.938	11.626	11.750	11.374	11.062	10.750	10.500	10.126	10.750
14	14.000	13.500	13.376	13.250	13.250	13.124	12.812	13.000	12.500	12.124	11.814	11.500	11.188	...
16	16.000	15.500	15.376	15.250	15.250	15.000	14.688	15.000	14.312	13.938	13.564	13.124	12.812	...
18	18.000	17.500	17.376	17.124	17.250	16.876	16.500	17.000	16.124	15.688	15.250	14.876	14.438	...
20	20.000	19.500	19.250	19.000	19.250	18.812	18.376	19.000	17.938	17.438	17.000	16.500	16.062	...
24	24.000	23.500	23.250	22.876	23.250	22.624	22.062	23.000	21.562	20.938	20.376	19.876	19.312	...
30	30.000	29.376	29.000	28.750	29.250	...	...	29.000	...	...	...	...	...	...
36	36.000	35.376	35.000	34.750	35.250	34.500	...	35.000	...	...	...	...	...	...
42	42.000	...	...	...	41.250	...	...	41.000	...	...	...	...	...	...

Fig. 5-10 Flange bore dimensions.

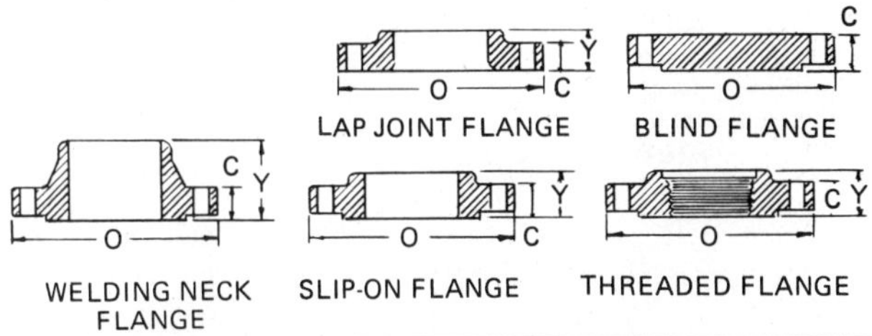
C
Y
O
C
O
LAP JOINT FLANGE
BLIND FLANGE
C
Y
O
C
Y
O
C
C
Y
O
WELDING NECK FLANGE
SLIP-ON FLANGE
THREADED FLANGE

Nom Pipe Size	150 LB. FLANGES O	C②	Y② Weld Neck	Y② Slip On Thrd.	Y② Lap Joint	Bolt Circle	No. and Size of Holes	300 LB. FLANGES O	C②	Y② Weld Neck	Y② Slip on Thrd.	Y② Lap Joint	Bolt Circle	No. and Size of Holes
½	3½	7/16	1⅞	⅝	⅝	2⅜	4-⅝	3¾	9/16	2 1/16	⅞	⅞	2⅝	4-⅝
¾	3⅞	½	2 1/16	⅝	⅝	2¾	4-⅝	4⅝	⅝	2¼	1	1	3¼	4-¾
1	4¼	9/16	2 3/16	11/16	11/16	3⅛	4-⅝	4⅞	11/16	2 7/16	1 1/16	1 1/16	3½	4-¾
1¼	4⅝	⅝	2¼	13/16	13/16	3½	4-⅝	5¼	¾	2 9/16	1 1/16	1 1/16	3⅞	4-¾
1½	5	11/16	2 7/16	⅞	⅞	3⅞	4-⅝	6⅛	13/16	2 11/16	1 3/16	1 3/16	4½	4-⅞
2	6	¾	2½	1	1	4¾	4-¾	6½	⅞	2¾	1 5/16	1 5/16	5	8-¾
2½	7	⅞	2¾	1⅛	1⅛	5½	4-¾	7½	1	3	1½	1½	5⅞	8-⅞
3	7½	15/16	2¾	1 3/16	1 3/16	6	4-¾	8¼	1⅛	3⅛	1 11/16	1 11/16	6⅝	8-⅞
3½	8½	15/16	2 13/16	1¼	1¼	7	8-¾	9	1 3/16	3 3/16	1¾	1¾	7¼	8-⅞
4	9	15/16	3	1 5/16	1 5/16	7½	8-¾	10	1¼	3⅜	1⅞	1⅞	7⅞	8-⅞
5	10	15/16	3½	1 7/16	1 7/16	8½	8-⅞	11	1⅜	3⅞	2	2	9¼	8-⅞
6	11	1	3½	1 9/16	1 9/16	9½	8-⅞	12½	1 7/16	3⅞	2 1/16	2 1/16	10⅝	12-⅞
8	13½	1⅛	4	1¾	1¾	11¾	8-⅞	15	1⅝	4⅜	2 7/16	2 7/16	13	12-1
10	16	1 3/16	4	1 15/16	1 15/16	14¼	12-1	17½	1⅞	4⅝	2⅝	3¾	15¼	16-1⅛
12	19	1¼	4½	2 3/16	2 3/16	17	12-1	20½	2	5⅛	2⅞	4	17¾	16-1¼
14	21	1⅜	5	2¼	3⅛	18¾	12-1⅛	23	2⅛	5⅝	3	4⅜	20¼	20-1¼
16	23½	1 7/16	5	2½	3 7/16	21¼	16-1⅛	25½	2¼	5¾	3¼	4¾	22½	20-1⅜
18	25	1 9/16	5½	2 11/16	3 13/16	22¾	16-1¼	28	2⅜	6¼	3½	5⅛	24¾	24-1⅜
20	27½	1 11/16	5 11/16	2⅞	4 1/16	25	20-1¼	30½	2½	6⅜	3¾	5½	27	24-1⅜
24	32	1⅞	6	3¼	4⅜	29½	20-1⅜	36	2¾	6⅝	4 3/16	6	32	24-1⅝
30	38¾	2⅛	5⅛	3½	..	36	28-1⅜	43	3⅝	8¼	8¼	..	39¼	28-1⅞
36	46	2⅜	5⅜	3¾	..	42¾	32-1⅝	50	4⅛	9½	9½	..	46	32-2⅛

Fig. 5-11 A portion of the flange rating chart.

very important, so it is very important that the pipe drafter be familiar with each.

Screwed or Slip-on Reducing Flange: This is used to reduce from one pipe size to another. The drilling of this flange corresponds to the drilling of a weld-neck flange (see Fig. 5-18, p. 43).

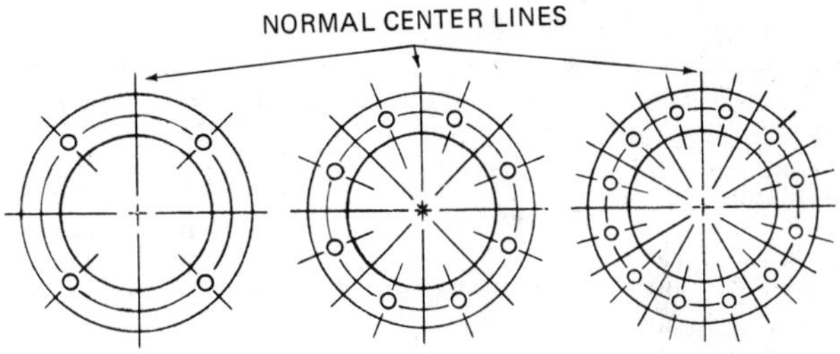
NORMAL CENTER LINES
FLANGE WITH 4 HOLES (STRADDLE CENTER LINES 90° APART)
FLANGE WITH 8 HOLES (STRADDLE CENTER LINES 45° APART)
FLANGE WITH 12 HOLES (STRADDLE CENTER LINES 30° APART)

Fig. 5-12 Bolt holes always straddle the center line unless noted otherwise.

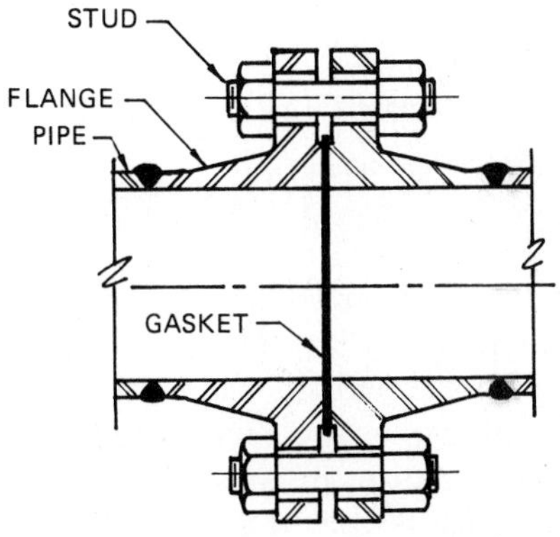
STUD
FLANGE
PIPE
GASKET

Fig. 5-13 Flange fit-up procedure.

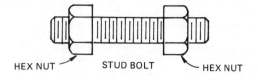
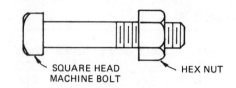

HEX NUT — STUD BOLT — HEX NUT

SQUARE HEAD MACHINE BOLT — HEX NUT

Nominal Pipe Size	150-lb FLANGES (1)				300-lb FLANGES (1)				400-lb FLANGES (2)			Nominal Pipe Size
	No. of Bolts or Studs	Dia. of Bolts or Studs	Length of Bolts A	Length of Studs B	No. of Bolts or Studs	Dia. of Bolts or Studs	Length of Bolts A	Length of Studs B	No. of Studs	Diam. of Studs	Length of Studs B	
½	4	½	1¾	2¼	4	½	2	2½	*	*	*	½
¾	4	½	2	2¼	4	⅝	2¼	3	*	*	*	¾
1	4	½	2	2½	4	⅝	2½	3	*	*	*	1
1¼	4	½	2¼	2½	4	⅝	2½	3¼	*	*	*	1¼
1½	4	½	2¼	2¾	4	¾	2¾	3½	*	*	*	1½
2	4	⅝	2½	3¼	8	⅝	2¾	3½	*	*	*	2
2½	4	⅝	2¾	3½	8	¾	3¼	4	*	*	*	2½
3	4	⅝	3	3½	8	¾	3½	4¼	*	*	*	3
3½	8	⅝	3	3½	8	¾	3½	4¼	*	*	*	3½
4	8	⅝	3	3½	8	¾	3¾	4½	8	⅞	5½	4
5	8	¾	3	3¾	8	¾	4	4¾	8	⅞	5¾	5
6	8	¾	3¼	4	12	¾	4	4¾	12	⅞	6	6
8	8	¾	3½	4¼	12	⅞	4½	5½	12	1	6¾	8
10	12	⅞	3¾	4¾	16	1	5¼	6¼	16	1⅛	7½	10
12	12	⅞	3¾	4¾	16	1⅛	5½	6¾	16	1¼	8	12
14	12	1	4¼	5¼	20	1⅛	5¾	7	20	1¼	8¼	14
16	16	1	4¼	5½	20	1¼	6¼	7½	20	1⅜	8¾	16
18	16	1⅛	4¾	6	24	1¼	6½	7¾	24	1⅜	9	18
20	20	1⅛	5	6¼	24	1¼	6¾	8¼	24	1½	9¾	20
24	20	1¼	5½	7	24	1½	7½	9¼	24	1¾	10¾	24

All dimensions are in inches. Lengths include allowance for pulling-up.
(1) Length based on ¹⁄₁₆″ raised, faced welding neck, slip-on, screwed, or blind flanges. For lap joint, add the thickness of both laps.

(2) Length based on ¼″ raised, faced welding neck, slip-on, screwed, or blind flanges. For lap joint, add the thickness of both laps and subtract ½″. For male-and-female or tongue-and-groove flange faces, subtract ¼″.

*Same as 600-lb flanges.

Fig. 5-14 Types of bolts and bolt lengths.

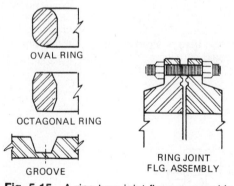

OVAL RING

OCTAGONAL RING

GROOVE

RING JOINT FLG. ASSEMBLY

Fig. 5-15 A ring-type joint flange assembly.

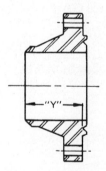

"Y"

Fig. 5-16 Weld-neck ring joint flange.

These flanges can be purchased with either concentric or eccentric reduction. The advantage of the eccentric threading is that it permits the bottom of the reduced section of pipe to be level with the large section of pipe, thus eliminating any possibility of a pocket. Another advantage is that it makes for an easier pipe-support problem. It is clear that with a concentric reduction in size in a reducing flange, a dam would be created. This dam would create a pocket in the larger size if the pipeline was horizontal. When a pocket of this nature cannot be avoided, it is good practice to use a drain valve so that in the event of shutdown the pocket can be efficiently drained.

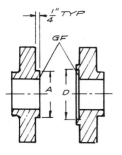 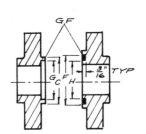 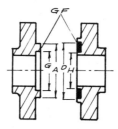 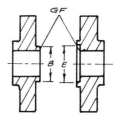

Bolt Lengths

| Nominal Pipe Size | OUTSIDE DIAMETER | | | | | | INSIDE DIAM. | |
| | Raised Face, Large Male, Large Tongue; also Pipe Lap | Small Male | Small Tongue | Large Female, Large Groove | Small Female | Small Groove | Large Tongue, Small Tongue | Large Groove, Small Groove |
	A	B	C	D	E	F	G	H
1/2	1 3/8	23/32	1 3/8	1 7/16	25/32	1 7/16	1	15/16
3/4	1 11/16	15/16	1 11/16	1 3/4	1	1 3/4	1 5/16	1 1/4
1	2	1 3/16	1 7/8	2 1/16	1 1/4	1 15/16	1 1/2	1 7/16
1 1/4	2 1/2	1 1/2	2 1/4	2 9/16	1 9/16	2 5/16	1 7/8	1 13/16
1 1/2	2 7/8	1 3/4	2 1/2	2 5/16	1 13/16	2 9/16	2 1/8	2 1/16
2	3 5/8	2 1/4	3 1/4	3 11/16	2 5/16	3 5/16	2 7/8	2 13/16
2 1/2	4 1/8	2 11/16	3 3/4	4 3/16	2 3/4	3 13/16	3 3/8	3 5/16
3	5	3 5/16	4 5/8	5 1/16	3 3/8	4 11/16	4 1/8	4 3/16
3 1/2	5 1/2	3 13/16	5 1/8	5 9/16	3 7/8	5 3/16	4 3/4	4 11/16
4	6 3/16	4 5/16	5 11/16	6 1/4	4 3/8	5 3/4	5 3/16	5 1/8
5	7 5/16	5 5/8	6 13/16	7 3/8	5 7/16	6 7/8	6 5/16	6 1/4
6	8 1/2	6 3/8	8	8 9/16	6 7/16	8 1/16	7 1/2	7 7/16
8	10 5/8	8 3/8	10	10 11/16	8 7/16	10 1/16	9 3/8	9 5/16
10	12 3/4	10 1/2	12	12 13/16	10 9/16	12 1/16	11 1/4	11 3/16
12	15	12 1/2	14 1/4	15 1/16	12 9/16	14 5/16	13 1/2	13 7/16
14	16 1/4	13 3/4	15 1/2	16 5/16	13 13/16	15 9/16	14 3/4	14 11/16
16	18 1/2	15 3/4	17 5/8	18 9/16	15 13/16	17 11/16	16 3/4	16 11/16
18	21	17 3/4	20 1/8	21 1/16	17 13/16	20 3/16	19 1/4	19 3/16
20	23	19 3/4	22	23 1/16	19 13/16	22 1/16	21	20 15/16
24	27 1/4	23 3/4	26 1/4	27 5/16	23 13/16	26 5/16	25 1/4	25 3/16

DIMENSIONS are shown in inches.

Fig. 5-17 Flange facings.

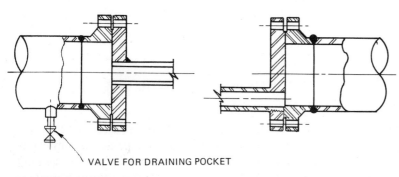

VALVE FOR DRAINING POCKET

CONCENTRIC SLIP-ON REDUCING FLANGE (NOTE POCKET)

ECCENTRIC SCREWED REDUCING FLANGE (NOTE ABSENCE OF POCKET)

Fig. 5-18 Reducing flanges.

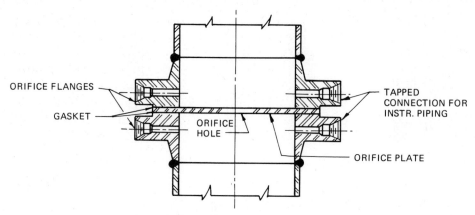

Fig. 5-19 Orifice flanges are used in pairs.

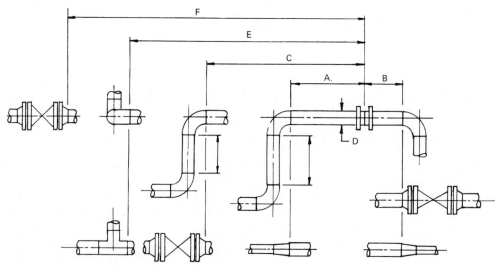

Fig. 5-20 Orifice flange runs in various situations.

Orifice Flange: This is used in pairs and in conjunction with an orifice plate, which is a disk with a hole that is usually considerably smaller in diameter than pipeline size (see Fig. 5-19). This disk is placed between the flanges. Note that the orifice flanges have two drilled and tapped holes on the side, 180° apart. The tapping of these holes varies with the size of the flange. Instrument piping is run from these connections to a meter, and flow is recorded by means of the differential in pressure on either side of the orifice plate. The location of these connections in the orifice flange is very important. It is customary to orient them so that they are on top in a gas line and on the side in a liquid line. Orifice flanges can be weld neck, screwed, or slip-on (see Fig. 5-19).

Jackscrews are also used with orifice flanges, for spreading the flanged joint to remove the orifice plate. The jackscrews are located 90° from the instrument taps (see Fig. 5-19).

Orifice flanges are ordered in pairs complete with taps and jackscrews. The length through the hub differs from that of regular flanges. The orifice plate between the flanges varies in thickness with the size of the pipe. The hole in the middle of the plate is sized to correspond to the rate of flow required.

The location of orifice flanges in a pipeline is very important to ensure a correct reading. They must be placed with a sufficient straight run before and after and with no interference such as branch connections, to ensure a flow with the least possible turbulence. They are normally located 30 diameters upstream and 5 diameters downstream of the flow in the pipe from ells, tees, or branch connections (see Fig. 5-20).

JACKSCREWS AND HOW THEY ARE USED

A jackscrew is installed by drilling and tapping only one of the companion flanges between any two bolts (see Fig. 5-21, p. 45). When it is necessary to swing the figure-8 blind from the one side to the other, and after all the bolts have been removed, the jackscrew is screwed in against the opposite companion flange which was not drilled and tapped.

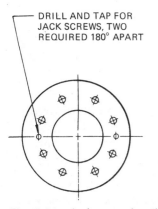

Fig. 5-21 Jack screw locations.

DRILL AND TAP FOR
JACK SCREWS, TWO
REQUIRED 180° APART

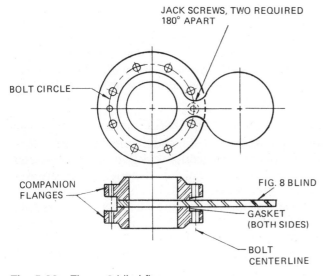

JACK SCREWS, TWO REQUIRED
180° APART

BOLT CIRCLE

COMPANION
FLANGES

FIG. 8 BLIND

GASKET
(BOTH SIDES)

BOLT
CENTERLINE

Fig. 5-22 Figure-8 blind flange.

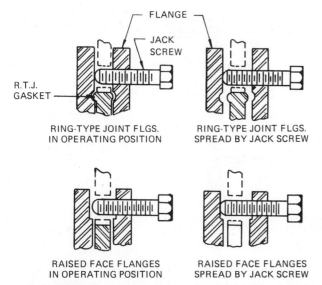

FLANGE

JACK
SCREW

R.T.J.
GASKET

RING-TYPE JOINT FLGS.
IN OPERATING POSITION

RING-TYPE JOINT FLGS.
SPREAD BY JACK SCREW

RAISED FACE FLANGES
IN OPERATING POSITION

RAISED FACE FLANGES
SPREAD BY JACK SCREW

Fig. 5-23 Jack screws in operating and spread position.

This forces the companion flanges apart, thus making it easy to swing the figure-8 blind using one jackscrew as the pivot (see Fig. 5-22, p. 45). Jackscrews are usually on flanges 4″ and above.

Figure 5-23 (p. 45) shows a typical detail of a jackscrew in the operating and spread position. Note also that it is imperative to use jackscrews when the flange facings are ring-joint or tongue-and-groove, because of the manner in which this type of facing creates an interlock of the flanges. Clearly, it would be impossible to swing a figure-8 blind flange unless a jackscrew were used to spread the flanges apart.

FIGURE-8 BLIND FOR FLANGED JOINTS

A figure-8 blind flange derives its name from its physical shape, which resembles the number 8. It is also known as a *spectacle blind*. A figure-8 blind flange is a combination of two disks. One disk has a hole with the diameter slightly larger than the inside diameter of the pipe and the bore of the flanges which make up the flanged joint. The other half of the figure-8 blind is a blank disk (no hole). With the blank half in the way of the flow, this flange offers a positive means of blanking off the flow and isolating a portion of a system. This is a safety measure for the protection of personnel when dismantling equipment or for other protective processes.

The figure-8 blind is installed between a pair of flanges with a gasket on both sides (see Fig. 5-22). The outside diameter of the figure 8 is such that it is slightly larger than the diameter of the raised face but is within the flange bolts. When a figure-8 blind flange is in place, an operator need only observe the disk which is exposed to determine whether the flow is on or off.

PADDLE BLIND FLANGE

A paddle blind flange derives its name from its physical shape, which resembles a paddle. A paddle blind is nothing more than a blind flange with a handle (see Fig. 5-24, p. 46). It is customary to use a spacer with this flange. The spacer is similar to the open half of the figure-8 blind (see Fig. 5-25, p. 46). It retains the gap necessary for the insertion of the paddle blind.

CAST-IRON FLANGES AND HOW THEY COMPARE TO FORGED-STEEL SIZES

In the 125- and 250-lb class of flanges previously mentioned, we are dealing with cast iron. These flanges are not welded to pipe, but are either furnished with threads in order that they may be attached to a threaded end of pipe or fitting or precast on the end of cast-iron pipe.

The outside diameter, the bolt circle, and the number of bolts in the 125-lb class are identical to the drilling of a 150-lb forged-steel weld-neck flange, but the thickness of the flange itself is different.

On the 125-lb cast-iron flange there is no raised face. This flange comes flat-faced, and is to be mated only with another flat-face flange.

The drilling of a 250-lb cast-iron flange is identical to the drilling of a 300-lb forged-steel flange. Only the thickness of the flanges differs.

The 250-lb cast-iron flange is fabricated with a 1/16" raised face.

Clearly, only flanges with similar drillings can be mated. That is, a 150-lb flange will mate with another 150-lb or a 125-lb flange, and a 300-lb flange will mate with another 300-lb or a 250-lb flange. When mating a steel flange with a raised face against a flat-face cast-iron flange on a piece of equipment, it is common practice to grind the raised face off the steel flange to prevent the cracking of the cast-iron flange when bolting pressure is applied.

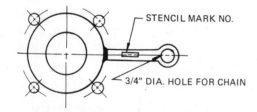

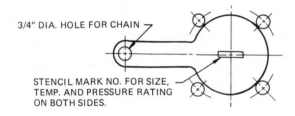

Fig. 5-24 Paddle blind spacer detail for raised face flanged joints.

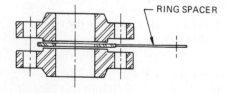

Fig. 5-25 Ring spacer for raised face and flanged joints.

FLANGE FINISHES

There are many finishes in general use for flange and lap joint facings. The finishes most commonly used are serrated and smooth.

Serrated Finish: The machining of faces with this finish is accomplished with a roundnose or form tool which cuts concentric grooves. On cast-iron and ferrosteel flange facings, there are approximately 16 serrations to the inch, and on steel, 32 serrations to the inch (see Fig. 5-26, p. 46).

Smooth Finish: The machining of faces with this finish is accomplished with tools which leave an even, uniform surface, designated as *smooth*.

Caution: Although not used commonly, the term *plane face* is sometimes encountered. It usually means smooth finish. Since it does not refer to a facing and is often confused with *plain* (or *flat*) *face*, the term is misleading and should be avoided (see Fig. 5-27, p. 46).

Special Finish (125 RMS Finish):   For services involving volatile, toxic, and flammable materials, a fine finish not exceeding 0.005" in depth between lands is sometimes used, depending on the designer's preference. Stock finish facing is a continuous spiral groove generated by a 1/16-in radius round-nose tool (1/8-in radius for sizes over 12") at a feed of 1/32 in (3/64 in for over 12"). See Fig. 5-28 (p. 46) for an enlarged drawing illustrating the developed surface for a 12" flange or smaller.

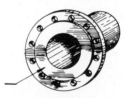

Fig. 5-26 Serrated finish flange.

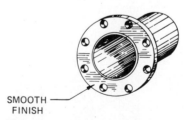

Fig. 5-27 Smooth finish flange.

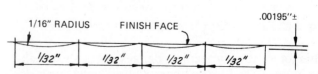

Fig. 5-28 Stock finish facing for a 12' flange.

Note: At present most flange manufacturers furnish either a "phonographic" or a concentric serrated facing as their general-purpose finish.

Finish practices for iron and steel flanged joints are as follows:

- Flanged joints when using 125-lb cast-iron or ferrosteel screwed flanges, are regularly furnished plain-faced with smooth finish. Serrated or special finish can be furnished, but at an extra charge.
- Flanged joints when using 250-lb cast-iron or ferrosteel screwed flanges, and 150- or 300-lb forged-steel screwed, weld-neck, or slip-on welding flanges, are regularly furnished with a $\frac{1}{16}$" high raised face having a serrated finish.
- Raised faces should be removed from 150-lb steel flanges before bolting them against 125-lb iron flanges, and only full-face gaskets should be used.
- Flanged joints utilizing steel valves and fittings with 400-, 600-, 900-, 1,500-, or 2,500-lb forged-steel screwed, weld-neck, or slip-on welding flanges bolted on are regularly furnished with a $\frac{1}{4}$" high raised face with serrated finish.

MANWAY COVERS

Manway covers are blind flanges, usually 18 in diameter and over. They are used to blind manways, which are a means of entrance into a vessel for the maintenance of the internals. Due to their weight and their frequent use, they are usually handled by means of a davit or hinge. There are three basic types of manway covers:

A. The manway hinge type (see Fig. 5-29)

B. The davit for horizontal manway (see Fig. 5-30)

C. The davit for vertical manway (see Fig. 5-31).

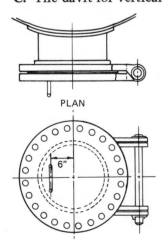

PLAN

ELEVATION

Fig. 5-29 Manway hinge-type detail.

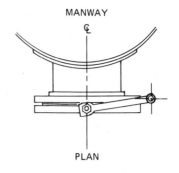

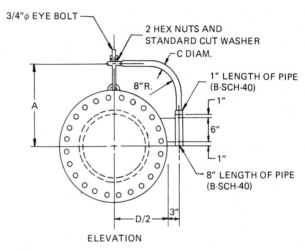

Fig. 5-30 Davit detail for horizontal manway.

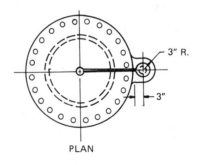

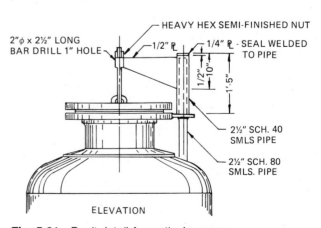

Fig. 5-31 Davit detail for vertical manway.

EXERCISES

5-1. Define the following terms:

1. Flange
2. Reference
3. Standard
4. Forged-steel flange
5. Cast-iron flange
6. Seal weld
7. Screwed flange
8. Weld-neck flange
9. Slip-on flange
10. Socket-weld flange
11. Blind flange or blank flange
12. Lap-stub-end flange
13. Facing
14. Stub end
15. Raised face
16. Flat face
17. Bore
18. Nominal size
19. Schedule
20. Bolt diameter
21. Machine bolt
22. Stud bolt
23. Finish
24. Ring-joint facing
25. Tongue-and-groove facing
26. Reducing flange
27. Concentric
28. Eccentric
29. Figure-8 blind flange
30. Jackscrew
31. Gasket
32. Paddle-blind flange
33. Ring spacer
34. Orifice flanges
35. Manway cover
36. Davit

5-2. Match the terms and abbreviations:

1. flg.
2. ref.
3. std.
4. scrd.

a. Standard
b. Cast iron
c. Screwed
d. Reference

5. C.I.
6. W.N.
7. S.O.
8. S.W.
9. B.F.
10. F.F.
11. R.F.
12. nom.
13. sch.
14. L.W.
15. B.C.
16. R.T.J.
17. red.
18. stl.
19. T.E.
20. wt.
21. X-strong
22. XX-strong
23. conc.
24. ecc.
25. gskt.

e. Slip-on
f. Socket weld
g. Flange
h. Blind flange
i. Flat face
j. Weld neck
k. Nominal
l. Lap weld
m. Raised face
n. Ring-type joint
o. Schedule
p. Thread end
q. Reducer
r. Bolt circle
s. Weight
t. Steel
u. Extra stong
v. Double extra strong
w. Concentric
x. Eccentric
y. Gasket

5-3. Write a brief report on how flanges are made.

5-4. Use the flange rating chart (Fig. 5-8) to look up the following dimensions:

1. O.D. of a 12″ pipe
2. Flange bore of schedule 40 nom. 8″ pipe
3. Y dimension of 900# weld-neck flange 24″
4. 0 dimension of 150#, 4″
5. Bolt circle diameter of a 2,500#, 12″ flange

5-5. Use the chart for types of bolts and bolt lengths (Fig. 5-14) to look up the following dimensions:

1. Machine blot length of a 10″ 150# flange rating
2. Stud bolt length of a 24″ 400# flange rating.

DRAWING PROBLEMS

5-6. Make a drawing to scale of a 6″, 600# slip-on flange, front and side view (use dimensions from chart).

5-7. Make a drawing to scale of an 8″, 2,500# blind flange, front and side view.

5-8. Make a detail drawing of two of the flanges in the system. Select scale and draw double line.

REVIEW QUESTIONS

5-8. List three types of manway covers.

5-9. Describe how jackscrews are located on flanges.

5-10. Describe what orifice flanges are and how they operate.

5-11. Explain the difference between machine bolts and stud bolts.

5-12. List the seven pressure ratings of forged-steel flanges.

5-13. Why should all bolt holes straddle centerlines?

5-14. List five different types of flanges.

5-15. Explain why flanges are used.

5-16. List two types of flange faces.

5-17. List two types of flange finishes.

Six

Pipe Fittings

There are many types and sizes of fittings, each having a particular application. Process piping uses primarily butt-welded, flanged, socket, and threaded connections. Occasionally, however, the bell-and-spigot is used for soil and sanitary lines. Soldered connections are very low pressure and low temperature and are seldom used for anything except some types of utility line. This chapter concentrates primarily on the major types of fittings and how they are used within a piping process system.

After completion of this chapter the student should be able to identify the major types of welded, flanged, and threaded fittings and know the particular application of each. Fitting charts and standard reference books will be used to reinforce this emphasis on pipe fittings.

PIPE FITTINGS AND HOW THEY ARE CONNECTED

A *fitting* is any type of pipe connector that is used to change size or direction, to make a connection, or to change the pipe specification from one point to another. There are dozens of fittings, and in this chapter we discuss the most common ones. Six basic methods are used to connect pipe and fittings, each of which has specific applications: welded, threaded, flanged, soldered, bell-and-spigot, and socket (see Fig. 6-1).

WELDED CONNECTIONS

A welded joint is safer and more economical than a flanged joint. As mentioned in Chap. 5, a flanged joint requires two flanges, bolts, a gasket, and also two circumferential welds, whereas a welded joint requires only one circumferential weld. Clearly, it is more economical. A welded joint is also safer, because it is leakproof; therefore, there is no danger, as with a flanged joint, of a gasket blowing and a hot or volatile material spraying, endangering lives, and creating a fire hazard. As for strength, a welded connection is as strong as or stronger than the pipe itself.

It is also well to remember that it is extremely costly if a refinery has to curtail or completely shut

down because of an uncalled-for, preventable maintenance problem. In the economics of good engineering, the cost of upkeep must be weighed against the initial cost. Refineries are now being designed with a minimum of flanged joints thereby making welded fittings a necessity.

Notice that the wall thickness and the inside diameter of a fitting are the same as those of pipe in the same classification. This common classification is called *schedule* (see Fig. 5-10). Fittings come in numerous types, sizes, and shapes. The butt-weld type is most commonly used in sizes 2″ and above. Similar to steel pipe, the butt-weld fittings most commonly used are seamless and are manufactured from carbon steel or special alloys. The manufacturing of seamless steel fittings is different from that of seamless steel pipe in that the fittings are forged.

There are several types of fittings, and each of them will be discussed, beginning with the several kinds of weld fittings (see Fig. 6-2).

BUTT-WELD ELBOWS

An *elbow*, or *ell* as it is commonly called, is a fitting used to change the direction of flow. The standard butt-weld ells available are for 180°, 90°, and 45° changes of direction. The 180° ell is also referred to as a *return bend*. All butt-weld ells are manufactured with beveled ends.

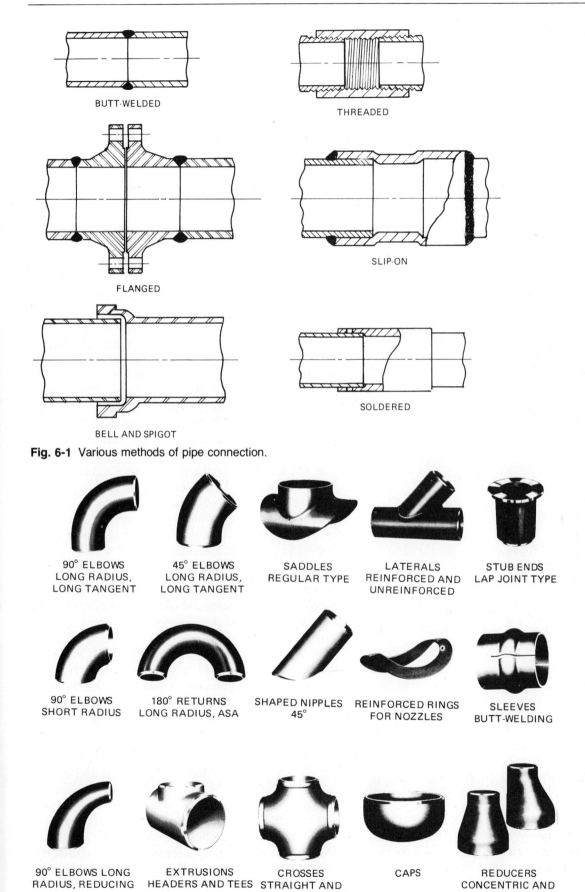

BUTT-WELDED

THREADED

FLANGED

SLIP-ON

BELL AND SPIGOT

SOLDERED

Fig. 6-1 Various methods of pipe connection.

90° ELBOWS
LONG RADIUS,
LONG TANGENT

45° ELBOWS
LONG RADIUS,
LONG TANGENT

SADDLES
REGULAR TYPE

LATERALS
REINFORCED AND
UNREINFORCED

STUB ENDS
LAP JOINT TYPE

90° ELBOWS
SHORT RADIUS

180° RETURNS
LONG RADIUS, ASA

SHAPED NIPPLES
45°

REINFORCED RINGS
FOR NOZZLES

SLEEVES
BUTT-WELDING

90° ELBOWS LONG
RADIUS, REDUCING

EXTRUSIONS
HEADERS AND TEES

CROSSES
STRAIGHT AND
REDUCING

CAPS

REDUCERS
CONCENTRIC AND
ECCENTRIC

Fig. 6-2 Standard weld fittings.

(a)

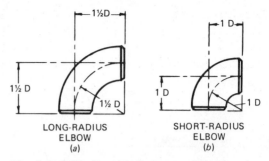

Fig. 6-4 90° butt-weld ells are available in a long radius *(a)* and a short radius *(b)*.

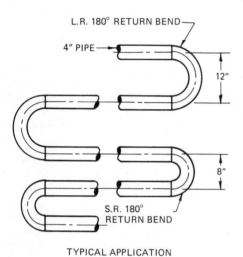

TYPICAL APPLICATION

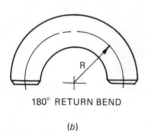

180° RETURN BEND

(b)

Fig. 6-3a A 180° elbow.

Fig. 6-3b Types of return elbows and applications.

180° Return Bend:

The 180° return bend (see Fig.6-3a) is available in short or long radius. In the long radius, r is equal to 1½ times the nominal pipe diameter. For example, a 4" long-radius return bend has a radius of 6". In the short radius, r is equal to the nominal pipe diameter. For example, a 4" short-radius return bend has a radius of 4". This fitting is most commonly used in the fabrication of coils (see Fig. 6-3b).

90° Butt-weld Ell:

The 90° butt-weld ell which is used to make a 90° offset in any direction (see Fig. 6-4) is also available in short and long radius. Long radius equals 1½ times nominal pipe diameter and short radius equals nominal pipe diameter. The long-radius ell is most commonly used. The short-radius ell is used only when the piping layout is so confined that only a short radius will solve the problem (see Fig. 6-5).

45° Butt-weld Ell:

The 45° butt-weld ell which is used to make a 45° offset in the piping (see Fig. 6-6) is available only in long radius; that is, r is equal to 1½ times the nominal pipe diameter. When an offset is required where the angle is more or less than the standard fitting, it is customary to take a 90° or a 45° fitting and cut it to the desired angle. When this is required it is standard practice to note it on the drawing (see Fig. 6-7). Such an elbow is called a *cutback elbow*. Cutback elbows are very commonly used, but require very careful calculations in order that the angles desired are accurate. Cutback charts and tables are used to determine cutback arc length (see Fig. 6-8). If the nominal pipe diameter and the number of degrees are known, the outside arc length as well as the inside arc length can be determined. For other cutback dimensions, see Fig. 6-9 (p. 54). This dimension is critical for dimensioning.

Mitred Elbow:

Another design which is employed to change direction of flow is a *mitred ell*. A mitred ell is fabricated from two or more straight lengths of pipe, properly cut and welded together to form the desired turn (see Fig. 6-10, p. 54). There are several basic types of mitred ells (see Fig. 6-11, p. 54). However, mitred ells are not very common because of the turbulence they create in the flow. Their advantage is their relatively economical quality in relation to a bent pipe or a special pipe fitting.

Reducing Welding Elbow:

A reducing welding ell is a substitute for a 90° weld ell and a weld reducer (see Fig. 6-12). This fitting is well tapered to give the best possible flow conditions. When referring to it, mention the larger diameter first—for example, *an 8" × 6" reducing weld ell* (see Fig. 6-13). This fitting is available in long radius only. Clearly, use of this fitting always saves the distance required for a weld reducer. It is a space saver and is used when space is minimum. It also has an economic advantage because it reduces the cost of a fitting and

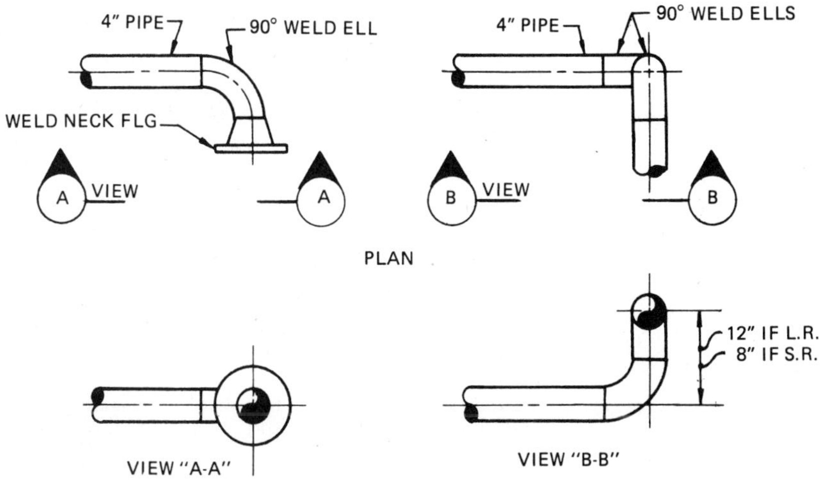

Fig. 6-5 90° butt-weld ell application.

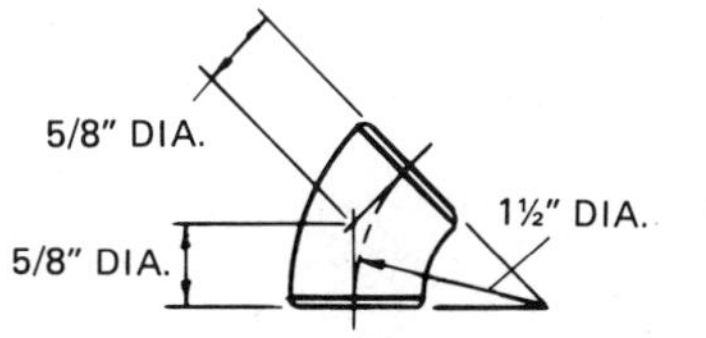

Fig. 6-6 A 45° butt-weld ell.

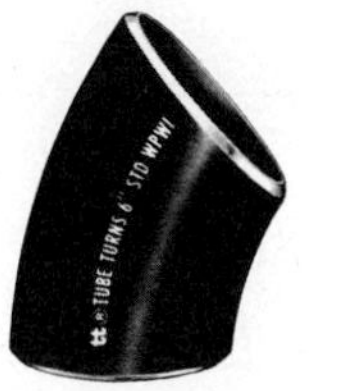

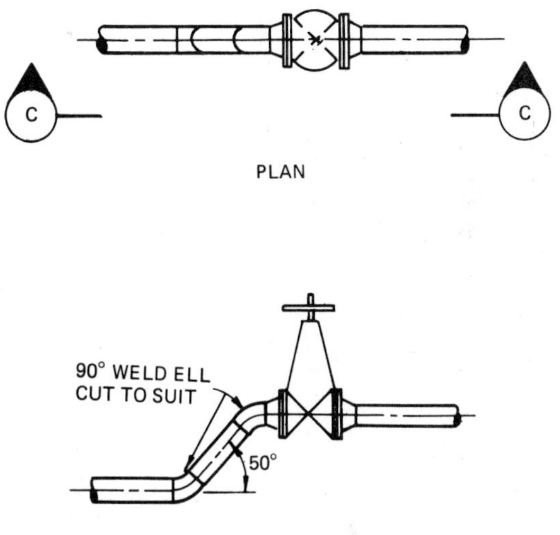

Fig. 6-7 Cutback elbow applications.

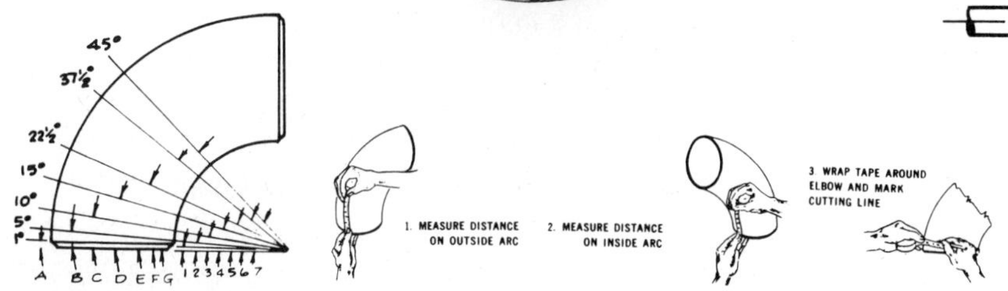

ODD DEGREE LONG RADIUS ELBOWS							
NOM SIZE	INSIDE ARC						
	1	2	3	4	5	6	7
2	1/32	5/32	5/16	15/32	23/32	1 3/16	1 7/16
2½	3/64	3/16	13/32	19/32	29/32	1½	1 13/16
3	3/64	¼	½	23/32	1 3/32	1 13/16	2 5/32
3½	1/16	9/32	9/16	27/32	1 9/32	2⅛	2 9/16
4	1/16	5/16	21/32	31/32	1 15/32	2 7/16	2 15/16
5	5/64	13/32	13/16	1¼	1 27/32	3 3/32	3 23/32
6	3/32	½	1	1½	2 7/32	3 23/32	4 15/32
8	⅛	11/16	1 11/32	2	3 1/32	5 1/32	6 1/32
10	5/32	27/32	1 11/16	2 17/32	3 25/32	6 5/16	7 9/16
12	7/32	1	2 1/32	3 1/16	4 9/16	7 19/32	9⅛
14	¼	1 7/32	2 7/16	3 21/32	5½	9 5/32	11
16	9/32	1 13/32	2 13/16	4 3/16	6 9/32	10 15/32	12⅝
18	5/16	1 9/16	3⅛	4 23/32	7 1/16	11 25/32	14⅛
20	11/32	1¾	3½	5¼	7 27/32	13 3/32	15 11/16
22	⅜	1 29/32	3 27/32	5¾	8⅝	14⅜	17 9/32
24	13/32	2 3/32	4 3/16	6 9/32	9 7/16	15 11/16	18 27/32
26	15/32	2 9/32	4 17/32	6 13/16	10 7/32	17 1/32	20 13/32
30	17/32	2⅝	5¼	7⅞	11 25/32	19⅝	23 9/16
34	19/32	2 31/32	5 29/32	8 29/32	13⅜	22 9/32	26 11/16
36	⅝	2 13/16	6¼	9 7/16	14⅛	23⅝	28¼
42	23/32	3 21/32	7 5/16	10 19/32	16½	26⅜	32 31/32

ODD DEGREE LONG RADIUS ELBOWS							
NOM. SIZE	OUTSIDE ARC						
	A	B	C	D	E	F	G
2	5/64	⅜	23/32	1 3/32	1 21/32	2¾	3 9/32
2½	3/32	7/16	29/32	1 11/32	2 1/32	3⅜	4 1/16
3	7/64	9/16	1⅛	1⅝	2 15/32	4 3/32	4 29/32
3½	⅛	⅝	1 9/32	1 29/32	2 27/32	4¾	5 11/16
4	9/64	23/32	1 7/16	2 5/32	3¼	5 13/32	6 15/32
5	3/16	29/32	1 25/32	2 11/16	4 1/32	6 23/32	8 1/16
6	7/32	1 1/16	2 5/32	3 7/32	4 27/32	8 1/16	9 21/32
8	9/32	1 7/16	2 27/32	4 9/32	6 13/32	10 11/16	12 13/16
10	11/32	1 25/32	3 9/16	5 11/32	8	13 11/32	16
12	7/16	2⅛	4¼	6⅜	9 9/16	15 31/32	19 5/32
14	½	2 7/16	4⅞	7 5/16	11	18 5/16	22
16	9/16	2 13/16	5 19/32	8⅜	12 9/16	20 15/16	25⅛
18	⅝	3⅛	6 9/32	9 7/16	14⅛	23 9/16	28 9/32
20	11/16	3½	7	10 15/32	15 23/32	26 3/16	31 13/32
22	¾	3 27/32	7 11/16	11 17/32	17 9/32	28 13/16	34 9/16
24	27/32	4 3/16	8⅜	12 9/16	18 27/32	31 13/32	37 11/16
26	29/32	4 17/32	9 3/32	13⅝	20 13/32	34 1/32	40 27/32
30	1 1/32	5¼	10 15/32	15¾	23 9/16	39¼	47⅛
34	1 5/32	5 29/32	11 27/32	17 13/16	26 23/32	44 17/32	53⅜
36	1 7/32	6¼	12 17/32	18⅞	28 7/32	47	56 17/32
42	1 7/16	7 5/16	14⅝	22	32 31/32	54 31/32	65 15/16

Fig. 6-8 Arc lengths for cutback elbows.

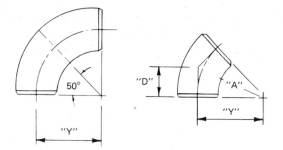

"D" (DISTANCE = (SINE 1/2 ANGLE "A") x LENGTH SIDE "Y")

Fig. 6-9 Cutback angle calculations.

Fig. 6-10 A mitred elbow is designed to change the direction of flow.

ONE (1) WELD MITRED 90° ELL

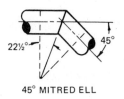

45° MITRED ELL

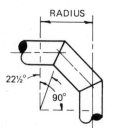

THREE (3)-PIECE MITRED 90° ELL

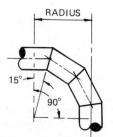

FOUR (4)-PIECE MITRED 90° ELL

Fig. 6-11 Basic types of mitred ells.

54

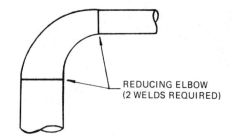

REDUCING ELBOW (2 WELDS REQUIRED)

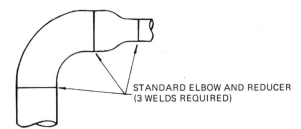

STANDARD ELBOW AND REDUCER (3 WELDS REQUIRED)

Fig. 6-12 Comparison of reducing ell with standard ell and weld reducer.

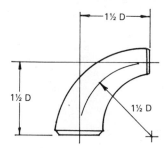

Fig. 6-13 90° reducing weld ell.

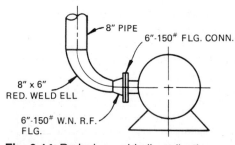

Fig. 6-14 Reducing weld ell application.

a circumferential weld. Its disadvantage is that it is not adaptable to other uses. When a company is stocking fittings, a reducing weld ell is not as flexible as a standard 90° ell. A practical application for a reducing ell is seen in Fig. 6-14.

WELDING TEE

When it is necessary to take a branch connection off a pipeline, 90° to the run, a butt-weld tee is used. Just as the name implies, a *tee* is a fitting with three openings forming a T (see Fig. 6-15). There are two commonly used types of butt-weld tees, a straight tee and a reducing tee.

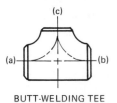

BUTT-WELDING TEE

Fig. 6-15 Butt-welding tee.

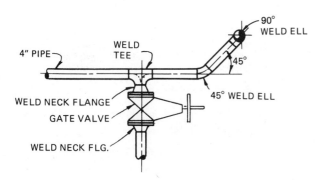

Fig. 6-16 Butt-weld straight tee.

ing cross, where the run a to b is larger than c to d, but c and d are the same size. For example, a reducing cross should be 6″ × 4″, or 6″ × 6″ × 4″ × 4″.

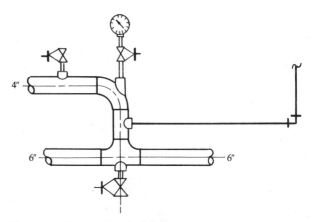

Fig. 6-17 Butt-weld reducing tee.

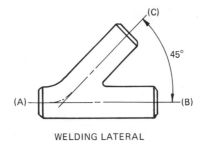

WELDING LATERAL

Fig. 6-18 Welding lateral.

Straight Tee: A straight tee is one whose three connections are all the same size. A 6″ straight tee is also referred to as 6″ × 6″ × 6″, or *6″ on the run and 6″ on the outlet.* The run sizes are the first two dimensions (a and b), and the outlet size is the third dimension (c) (see Fig. 6-15). All tee sizes are read in this manner (see Fig. 6-16 for application).

Reducing Tee: A reducing tee is one where the outlet is of a smaller size than the run. For example, if the pipe run size is 6″ and the branch is 4″, one would call for a 6″ × 6″ × 4″ butt-weld tee (see Fig. 6-17 for application).

MISCELLANEOUS BUTT-WELD FITTINGS

Lateral: A lateral is similar to a tee except that the outlet is 45° to the run (see Fig. 6-18). Laterals are also available in straight and reducing types. Reduction is only on the outlet, not on the run.

Butt-weld Cross: A butt-weld cross is available (see Fig. 6-19, p. 56), but not very often used. It is manufactured as a *straight-size cross* where a, b, c, and d are the same nominal pipe size; and as a *reduc-*

Welding Reducer: A butt-weld reducer is used when it is necessary to have a reduction in a pipe run (see Figs. 6-20 and 6-21, p. 56). When referring to a reducer, always mention the larger size first—for example, 8″ × 6″.

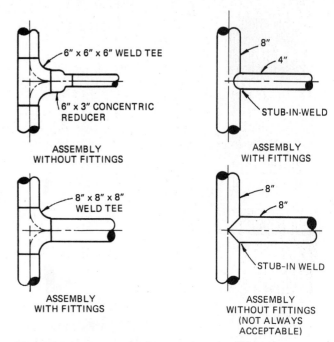

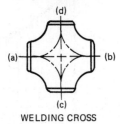

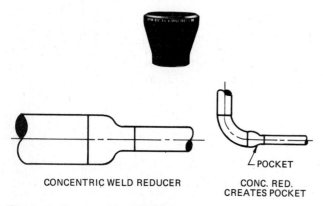

(d)

(a) (b)

(c)

WELDING CROSS

Fig. 6-19 Welding cross.

CONCENTRIC WELD REDUCER

POCKET

CONC. RED.
CREATES POCKET

Fig. 6-20 Concentric weld reducer.

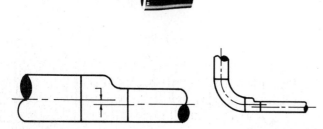

ECCENTRIC WELD REDUCER

ECC. RED. ELIMINATES
POCKET

Fig. 6-21 Eccentric weld reducer.

There are two types of reducers, concentric and eccentric. A concentric reducer maintains the same centerline straight through the large size and reduced size. An eccentric reducer does not, as is readily seen in Fig. 6-21. An eccentric reducer is used when the bottom of pipe is to be maintained level, for example, to eliminate a pocket (a place where liquid can remain stagnant), or keep all bottom ele-

6" x 6" x 6" WELD TEE

6" x 3" CONCENTRIC
REDUCER

ASSEMBLY
WITHOUT FITTINGS

8"

4"

STUB-IN-WELD

ASSEMBLY
WITH FITTINGS

8" x 8" x 8"
WELD TEE

ASSEMBLY
WITH FITTINGS

8"

8"

STUB-IN WELD

ASSEMBLY
WITHOUT FITTINGS
(NOT ALWAYS
ACCEPTABLE)

Fig. 6-22 Stub-in application.

vations equal for support purposes in a pipe rack (any large group of lines).

Butt Welding of Branch Connection: It is common practice with some companies today to economize on pipe fabrication by eliminating weld tees and reducers when possible.

This is done by the stub-in welding of pipe (see Fig. 6-22). This stub-in method is much more economical than the use of reducing tees. Some companies' standards limit the size of the branch which can be stub-in welded into a main line. When piping assemblies are being fabricated in a shop rather than in the field, it is possible to stub-in a branch of the same size as the main, although this practice is not acceptable to some companies. Same-size butt welding is more efficiently done in a pipe fabricator's shop rather than in the field.

Welding Cap: When it is necessary to blank off the end of a steel pipe, it can be done with flanges, as previously discussed under blind flanges (see Fig. 5-6), or with a butt-weld cap (see Fig. 6-23 *a, b,* and *c.*)

When blanking off a main pipeline which has possibilities of being extended in the future, remember that it is easier to remove a flange than to burn off a weld cap for the extension. A branch connection can be very easily butt-welded into a weld cap. It is often done for draining.

WELDS AND WELD SYMBOLS

Weld symbols are an abbreviated method of telling how and in what manner two pieces of metal are to be connected.

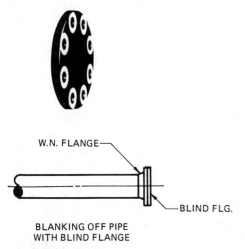

W.N. FLANGE

BLIND FLG.

BLANKING OFF PIPE
WITH BLIND FLANGE

Fig. 6-23 *a* Blind flange cap.

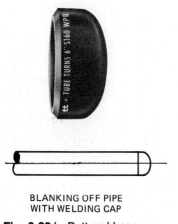

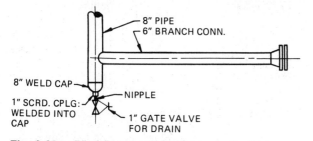

BLANKING OFF PIPE
WITH WELDING CAP

Fig. 6-23 *b* Butt-weld cap.

8" PIPE
6" BRANCH CONN.

8" WELD CAP
NIPPLE
1" SCRD. CPLG:
WELDED INTO
CAP
1" GATE VALVE
FOR DRAIN

Fig. 6-23 *c* Blind flange and weld cap applications.

Weld Symbol Standardization:
Weld symbols in pipe drafting are based on American Welding Society (AWS) specifications. Weld codes and specifications are becoming more and more important as new metals and alloys are beginning to be used. Without weld codes and specifications there would be no consistency in welds; some would be too thick, some would be too thin, and dissimilar metals welded together would create serious inconsistencies. Welding symbols within a particular company must be consistent with AWS specifications; that is, they must be equal to or better than AWS specs. Beyond that, each company has the freedom to develop its own specifications.

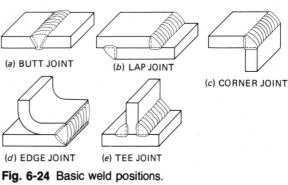

(a) BUTT JOINT   (b) LAP JOINT   (c) CORNER JOINT

(d) EDGE JOINT   (e) TEE JOINT

Fig. 6-24 Basic weld positions.

"V"   BEVEL   "U"   "J"

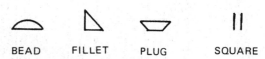

BEAD   FILLET   PLUG   SQUARE

Fig. 6-25 Edge preparation for welds.

There are many types of welds, and the study of weld symbols and specifications is quite involved. This text is only a brief introduction and does not go into lengthy explanation of each type of weld.

Basic Methods of Welding:
Resistance

Gas

Arc

Gas and arc are the most commonly used welding methods in piping and will be emphasized in this text.

Basic Weld Positions (see Fig. 6-24):
Corner

Butt

Lap

Tee

Edge

Methods of Edge Preparation (see Fig. 6-25):
Bead	V
Fillet	Bevel
Plug	U
Square	J

Standard Location of Elements of a Welding Symbol:
The student should become thoroughly familiar with this location diagram (Fig. 6-26, p. 58) and know the meaning of each element.

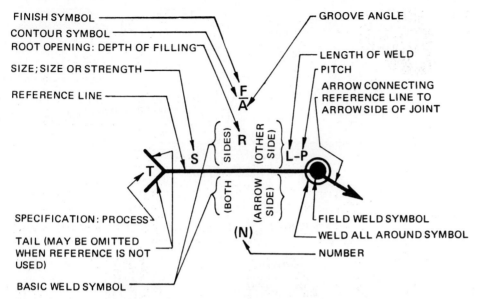

FINISH SYMBOL
CONTOUR SYMBOL
ROOT OPENING: DEPTH OF FILLING
SIZE; SIZE OR STRENGTH
REFERENCE LINE
GROOVE ANGLE
LENGTH OF WELD
PITCH
ARROW CONNECTING REFERENCE LINE TO ARROW SIDE OF JOINT
SPECIFICATION: PROCESS
TAIL (MAY BE OMITTED WHEN REFERENCE IS NOT USED)
BASIC WELD SYMBOL
FIELD WELD SYMBOL
WELD ALL AROUND SYMBOL
NUMBER

Fig. 6-26 Basic weld symbol nomenclature.

Arrow Side, Other Side, and Both Sides:

Figure 6-27 provides additional information about the location or position of the weld on fillet, U, and bevel welds. The principle is the same with the other types of welds. A complete up-to-date AWS chart can be obtained by writing to the American Welding Society.

THREADED PIPE FITTINGS

It is very common to find that pipe and fittings in sizes 1½" and below are threaded.

The classifications for screwed fittings are not the same as for welded fittings, and the materials are more numerous than for welded fittings. Below are listed the most common classifications and materials used for screwed fittings:

RATING	MATERIAL
2,000 lb, 3,000 lb, and 6,000 lb	Forged steel, carbon-molybdenum, stainless steel
150 lb and 300 lb	Malleable iron
125 lb and 250 lb	Cast iron, Brass

180° Screwed Return Bends:

These are available in three different designs: closed, medium, and open types. The dimensions a, b, and c differ for each of the three types (see Fig. 6-28). The closed type has the smallest dimension. The medium type is larger, and the open type is the largest. The variety of designs for screwed fittings is more numerous than for welded fittings.

90° Screwed Ells:

These elbows are available in straight sizes or in almost any reducing size with

LOCATION SIGNIFICANCE	FILLET	U	BEVEL
ARROW SIDE			
OTHER SIDE			
BOTH SIDES			

Fig. 6-27 Arrow side and other side or both sides.

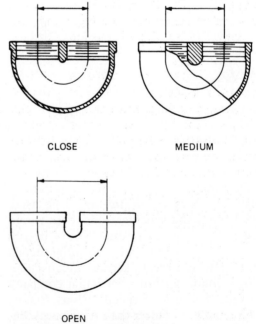

CLOSE

MEDIUM

OPEN

Fig. 6-28 Types of 180° screwed return bends—closed, medium, and open.

Fig. 6-29 Types of screwed elbows.

Fig. 6-30 45° straight elbow (left) and street ell (right).

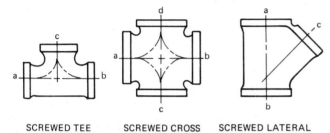

SCREWED TEE   SCREWED CROSS   SCREWED LATERAL

Fig. 6-31 Screwed tees, crosses, and laterals.

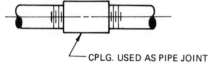

CPLG. USED AS PIPE JOINT

Fig. 6-32 Threaded coupling used as a pipe connector.

male ends. (see Fig. 6-29) An elbow with a male-and-female end is called a *street ell.*

45° Screwed Ell:
The 45° ell is available only in straight sizes and in street ells (see Fig. 6-30). Another important factor is that cutbacks are not possible with threaded fittings.

Screwed Tee, Cross, and Lateral:
These three fittings are available in almost any combination of reduction to suit the requirements of any piping layout. The tees are the most commonly used in this grouping. Here again, when referring to a reducing tee, lateral, or cross, the size of the run comes first before the outlet (see Fig. 6-31).

Screwed Coupling:
The screwed coupling has many uses besides being the connector between two lengths of threaded pipe (see Fig. 6-32). One very common use is to weld a threaded coupling into a steel pipe of a size larger than the coupling and use it to connect either an instrument, drain, vent, or branch line (see Fig. 6-33).

Screwed Reducer:
This is a screwed coupling with a reduction on one end. It is available in almost any combination of reductions.

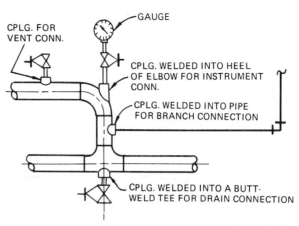

GAUGE

CPLG. FOR VENT CONN.

CPLG. WELDED INTO HEEL OF ELBOW FOR INSTRUMENT CONN.

CPLG. WELDED INTO PIPE FOR BRANCH CONNECTION

CPLG. WELDED INTO A BUTT-WELD TEE FOR DRAIN CONNECTION

Fig. 6-33 Threaded coupling used as a connector for vent, drain, and instruments.

Fig. 6-34 Screwed unions.

Screwed Union:
Unions play a very important part in a system where threaded pipe and fittings are used. A union is used as the makeup joint for easy disconnecting and also for easy assembling of a threaded pipe run.

A union is constructed of three parts, one union ring and two sleeves (see Fig. 6-34). The sleeves are assembled to the pipe, and the union ring pulls up to make a tight joint. In order to ensure a joint against leaking, the sleeves are ground or faced so that they seat tightly on each other. This union is called a *ground-seat union* and is most commonly used. There are two basic types, the female and the combination male and female.

Screwed Cap:
The *screwed cap* is used for blanking off the end of a threaded pipe or fitting. Caps are often used when a line is temporarily out of service or blanked temporarily for future service (see Fig. 6-35, p. 60).

Screwed Fittings, Conclusion:
There are many more types of screwed fittings listed in any

59

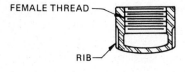

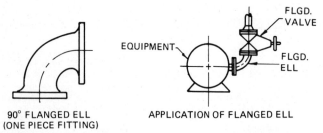

FEMALE THREAD

RIB

SCREWED CAP

Fig. 6-35 Screwed-cap—plain and flat band.

90° FLANGED ELL
(ONE PIECE FITTING)

APPLICATION OF FLANGED ELL

FLGD.
VALVE

EQUIPMENT

FLGD.
ELL

Fig. 6-36 Flanged fittings.

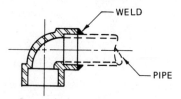

WELD

PIPE

90° SOCKET-WELDED ELL

Fig. 6-37 90° socket-welded ell.

manufacturer's catalog. However, note that the differences are usually variations of the ell, tee, and coupling with reductions and unions utilized in various ways. It is the responsibility of the drafter to become familiar with the basic ones covered in this chapter.

FLANGED FITTINGS

There are other types of fittings, but they are not used as commonly as the butt-weld and screwed types. One of these is the flanged fitting. The difference between a flanged fitting and a butt-weld fitting is just as the name implies: a flanged fitting is forged or cast with a flange on it which allows it to be bolted directly to a valve flange or another fitting. No weld is necessary (Fig. 6-36).

Flanged fittings are available in about the same variety as butt-weld fittings, such as 180°, 90°, and 45° straight-size ells; 90° reducing ells; straight and reducing tees; laterals; and reducers.

Flanged fittings have an economic value when they can be used to eliminate unnecessary welding. A typical example is shown in Fig. 6-36. Notice that a flange was required to mate the flange at the equipment and a flange was required to mate the flange on the valve. If one were to use a butt-weld ell with two weld-neck flanges, the cost of the flanges, fitting,

and welding would be more than that of the flanged fitting. However, some customers wish to hold flanged fittings to a minimum and prefer to standardize their stock with built-up elbows for maintenance.

In the interest of space, other types of flanged fittings are covered only in the Appendix.

SOCKET-WELD FITTINGS

Another fitting is the socket-weld type. It is welded in the same manner as the socket-weld flange which we previously reviewed. Socket-weld fittings are available in about the same variety as butt-weld fittings —90° ells, flanges, tees, etc. Pipe is inserted into the fitting until it butts against the shoulder, and then it is welded (see Fig. 6-37). Since there is such a similarity between threaded and socket-weld fittings, each basic type of socket weld has not been shown.

SPECIAL BONNEY WELD FITTINGS

Bonney weld fittings are welding fittings forged from steel of low carbon content. They are designed for making right-angle branch-pipe outlets such as tees, crosses, and side outlets by welding. They eliminate the preparation of preliminary layouts of main branch pipes. No templates are used, nor is any complicated cutting, threading, forming, and fitting of the main pipe necessary. Branch connections made with these fittings are strong, permanent, and leakproof.

Bonney weld fittings can be of three types: threaded outlet, welding outlet, and socket outlet. Bonney fittings come not only in Weldolets, but also in Sweepolets, Latrolets, Brazolets, etc.

Welding Outlet: A right-angle branch-pipe outlet can be made by welding a welding outlet to the main pipe. An ordinary V weld is used. The bottom opening is made after the fitting is attached to the main pipe. The junction is completed by welding the branch pipe to the outlet. Where the outlet saddles on the run pipe, it tapers at the proper angles and provides a single bevel groove joint at the crotch section, blending into a V-butt joint at the ear portion. (The abbreviation used in industry is W.O.L.; see Fig. 6-38.)

Threaded Outlet: Threaded outlets are installed on the main line in the same manner as welding outlets. A V weld is used. The threaded branch pipe is screwed into position in the tapped outlet of the fitting. Threaded outlets are widely used as radiator take offs, and in other locations where it is desirable to make connections without disturbing the main pipe. (The abbreviation used in industry is T.O.L.; see Fig. 6-39.)

Socket Outlets: Socket outlets are similar to the welding and threaded types except that the outlet is bored to accept standard outside pipe diameters. They are installed in the same manner as welding and threaded outlets. No beveling or threading of the branch pipe is necessary. (The abbreviation used in industry is S.O.L.; see Fig. 6-40.)

USE OF STANDARD SPECIFICATION SHEETS —WELDED, SCREWED, SOCKET

The principle of using a standard specification sheet is the same for all types of fittings. It is simply a matter of selecting the correct specification sheet, looking at the necessary information on the sheet, and interpreting the information accurately. The Appendix has a sample of standard specification sheets and a set of plans.

DELINEATION OF FITTINGS AND PIPING

Some companies have established standards for the manner in which piping drawings are to be drawn. All small-diameter pipe is generally drawn with a single line. Some companies single-line 4″ and below, but this depends on company standards

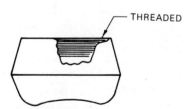

Fig. 6-38 Welding outlet.

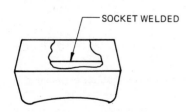

Fig. 6-39 Threaded outlet.

Fig. 6-40 Socket outlet.

and procedures. All large-diameter pipe is drawn with double lines. Here again, some companies double-line 4″ and above, but this also depends on company standards and procedures. Naturally all the fittings and valves must follow a similar pattern and conform to standard methods of representation.

Comparison of one line and two line for the butt-weld, screwed, and socket-weld fittings is shown in Fig. 6-41*a*, *b*, and *c* (pp. 61, 62).

Convenient reference charts for comparing single- to double-line fitting representations are provided in Figs. 6-42 (butt-weld), 6-43 (screwed) on p. 63, and 6-44, p. 64 (flanged). Figure 6-45, which begins on p. 65, is another convenient reference for some of the most common fittings in various positions.

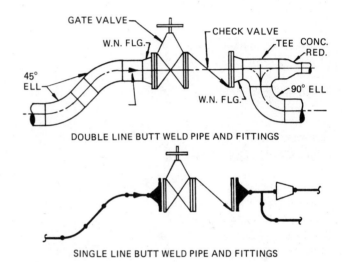

Fig. 6-41*a* One- and two-line comparison of butt-welded fittings.

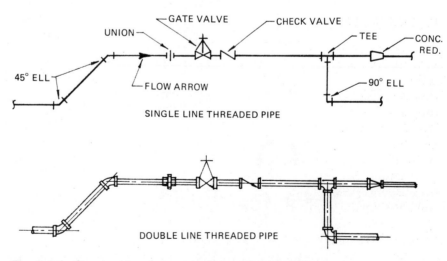

Fig. 6-41*b* One- and two-line comparison of threaded fittings.

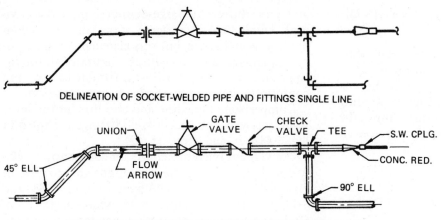

DELINEATION OF SOCKET-WELDED PIPE AND FITTINGS SINGLE LINE

DELINEATION OF SOCKET-WELDED PIPE AND FITTINGS DOUBLE LINE

Fig. 6-41c One- and two-line comparison of socket-weld fittings.

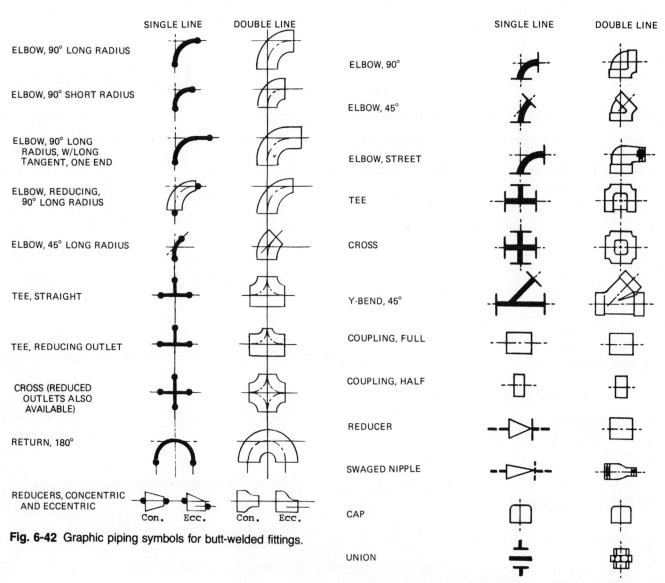

Fig. 6-42 Graphic piping symbols for butt-welded fittings.

Fig. 6-43 Graphic piping symbols for forged-steel screwed fittings.

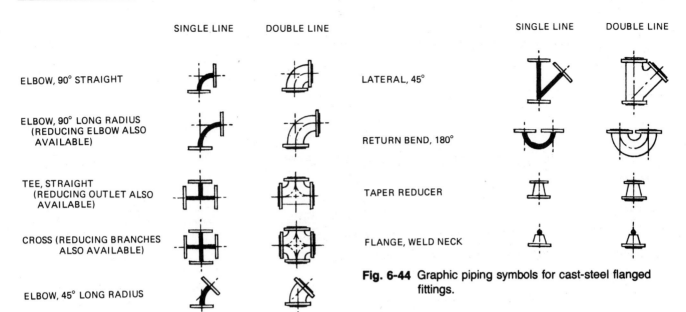

SINGLE LINE | DOUBLE LINE

ELBOW, 90° STRAIGHT

ELBOW, 90° LONG RADIUS (REDUCING ELBOW ALSO AVAILABLE)

TEE, STRAIGHT (REDUCING OUTLET ALSO AVAILABLE)

CROSS (REDUCING BRANCHES ALSO AVAILABLE)

ELBOW, 45° LONG RADIUS

SINGLE LINE | DOUBLE LINE

LATERAL, 45°

RETURN BEND, 180°

TAPER REDUCER

FLANGE, WELD NECK

Fig. 6-44 Graphic piping symbols for cast-steel flanged fittings.

WELD FITTINGS – SINGLE LINE

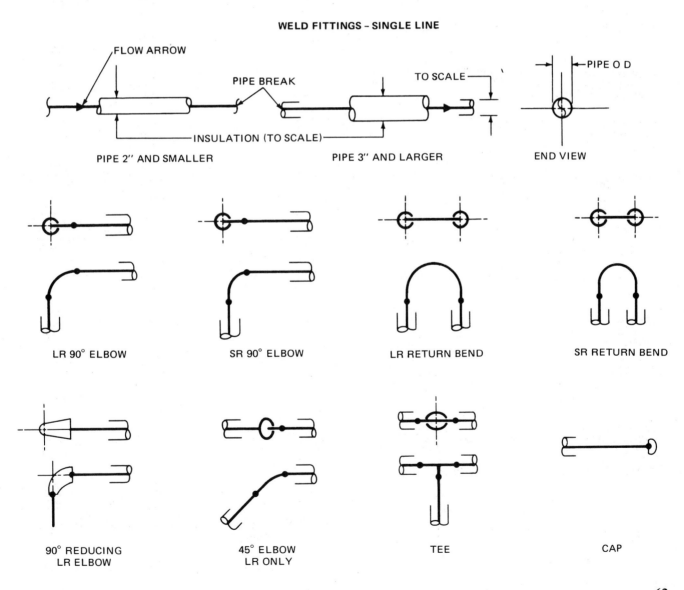

FLOW ARROW

PIPE BREAK

TO SCALE

PIPE O D

INSULATION (TO SCALE)

PIPE 2″ AND SMALLER

PIPE 3″ AND LARGER

END VIEW

LR 90° ELBOW

SR 90° ELBOW

LR RETURN BEND

SR RETURN BEND

90° REDUCING LR ELBOW

45° ELBOW LR ONLY

TEE

CAP

WELD FITTINGS – SINGLE LINE (continued)

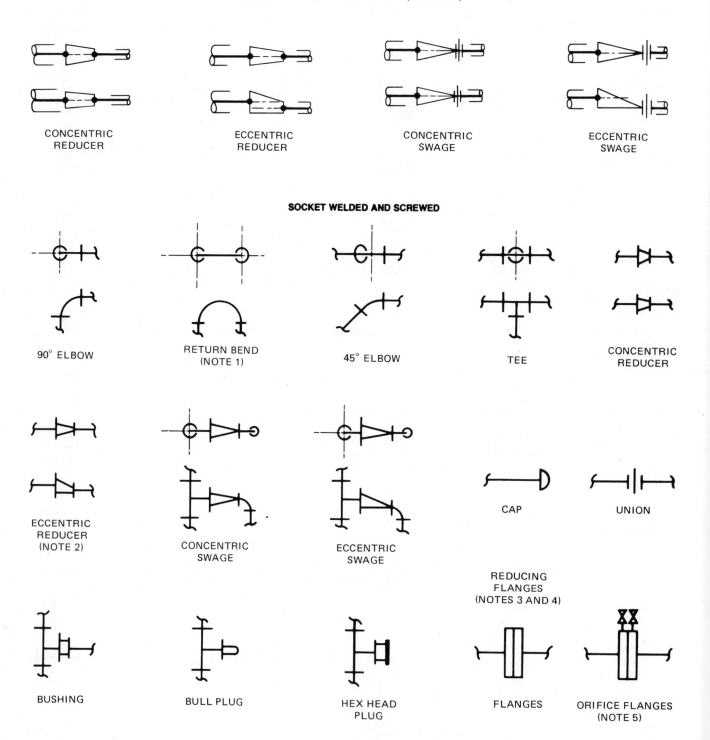

CONCENTRIC REDUCER ECCENTRIC REDUCER CONCENTRIC SWAGE ECCENTRIC SWAGE

SOCKET WELDED AND SCREWED

90° ELBOW RETURN BEND (NOTE 1) 45° ELBOW TEE CONCENTRIC REDUCER

ECCENTRIC REDUCER (NOTE 2) CONCENTRIC SWAGE ECCENTRIC SWAGE CAP UNION

REDUCING FLANGES (NOTES 3 AND 4)

BUSHING BULL PLUG HEX HEAD PLUG FLANGES ORIFICE FLANGES (NOTE 5)

NOTES:
1. RETURN BEND AVAILABLE IN CAST IRON AND MALLEABLE IRON, SCREWED FITTINGS ONLY.
2. ECCENTRIC REDUCER AVAILABLE IN CAST IRON, SCREWED FITTINGS ONLY.
3. EXAMPLE OF 4″ TO 2″ REDUCTION: 2″ X 9″ OD, 150 LBS, RF THREADED, REDUCING FLANGE.
4. SUBSTITUTE FOR REDUCING FLANGE: DRILL AND TAP BLIND FLANGE.
5. ORIFICE FLANGES PURCHASED IN PAIRS.

Fig. 6-45 Quick reference fitting-symbol chart.

SINGLE LINE VALVES

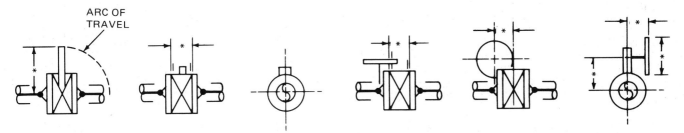

BUTTERFLY (WAFER, WRENCH OPERATED)   BUTTERFLY (WAFER, GEAR OPERATED)

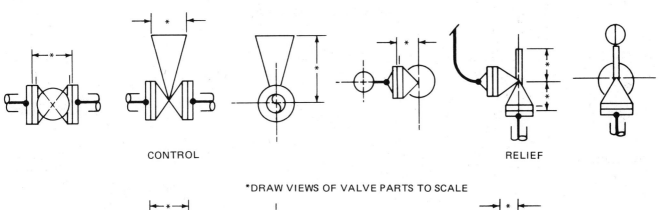

CONTROL   RELIEF

*DRAW VIEWS OF VALVE PARTS TO SCALE

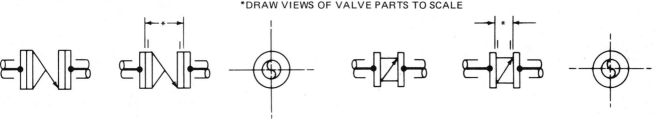

CHECK (SWING, LIFT, PISTON)   CHECK (WAFER)

VALVE END CONNECTIONS

DOUBLE LINE   SINGLE LINE

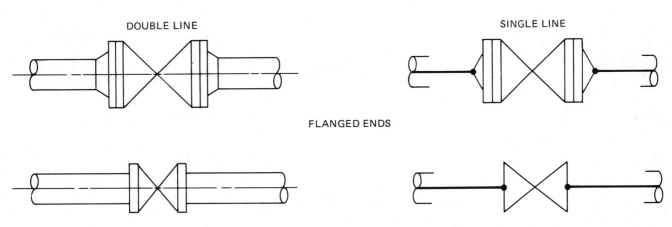

FLANGED ENDS

VALVE END CONNECTIONS (continued)

SCREWED AND SOCKET WELD ENDS

WELD ENDS

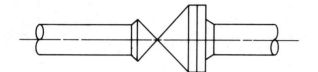 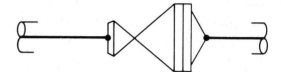

WELD END X FLANGED END

WELD FITTINGS – SINGLE LINE

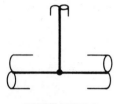 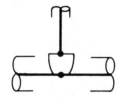 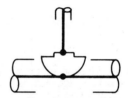 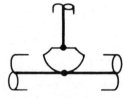

NOZZLE WELD   NOZZLE WELD
W/REINFORCING PAD   NOZZLE WELD
W/SADDLE   SWEEPOLET

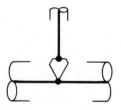 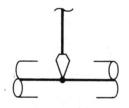 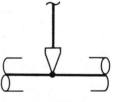 

BUTT WELDOLET   THREDOLET-
SOCKOLET   COUPLING   BUTT LATROLET

WELD FITTINGS – SINGLE LINE (continued)

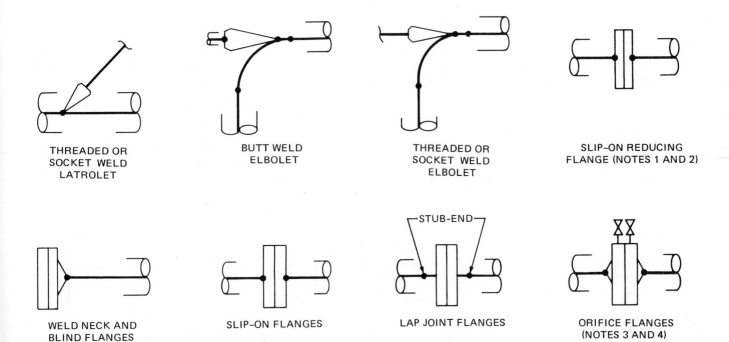

THREADED OR
SOCKET WELD
LATROLET

BUTT WELD
ELBOLET

THREADED OR
SOCKET WELD
ELBOLET

SLIP-ON REDUCING
FLANGE (NOTES 1 AND 2)

WELD NECK AND
BLIND FLANGES

SLIP-ON FLANGES

LAP JOINT FLANGES

ORIFICE FLANGES
(NOTES 3 AND 4)

NOTES:
1. REDUCING FLANGES ALSO AVAILABLE IN WELD NECK.
2. EXAMPLE OF 4" TO 2" REDUCTION: 2" X 9" OD, 150 RF, SO, RED, FLG.
3. ORIFICE FLANGES ALSO AVAILABLE IN SLIP-ON.
4. ORIFICE FLANGES PURCHASED IN PAIRS.

MISCELLANEOUS ARRANGEMENTS

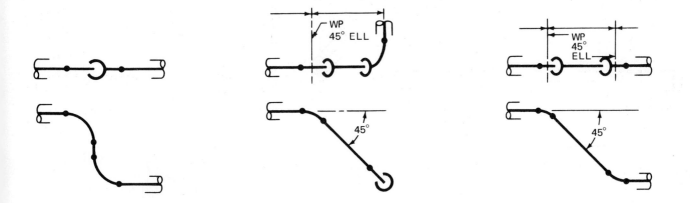

MISCELLANEOUS ARRANGEMENTS (continued)

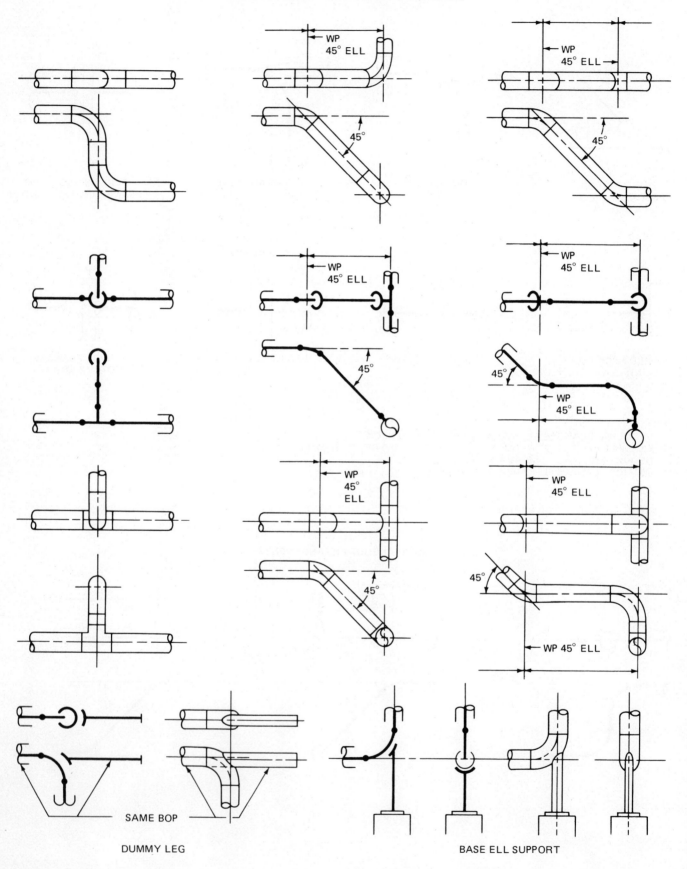

DUMMY LEG

BASE ELL SUPPORT

MISCELLANEOUS ARRANGEMENTS (continued)

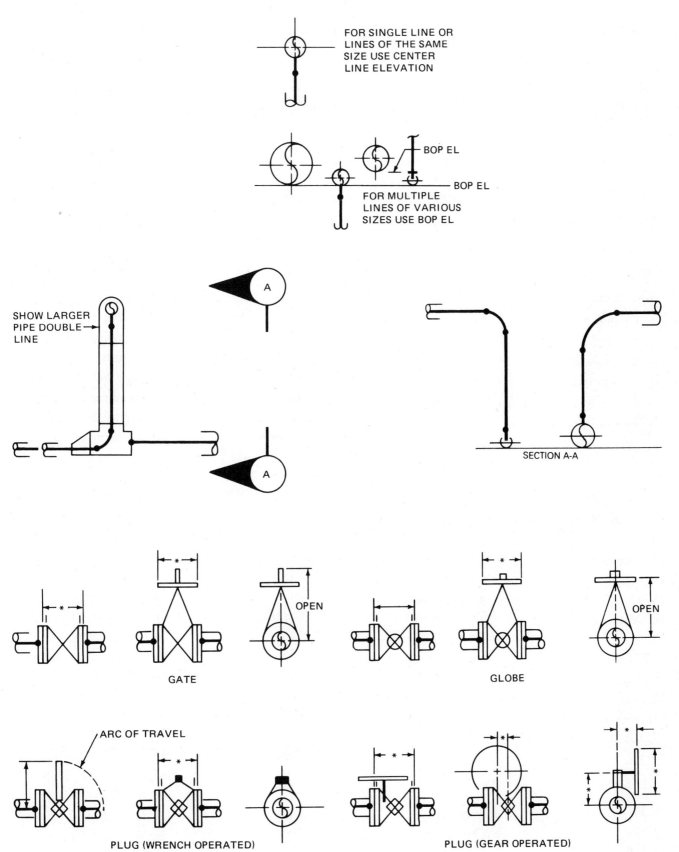

FOR SINGLE LINE OR LINES OF THE SAME SIZE USE CENTER LINE ELEVATION

BOP EL

BOP EL

FOR MULTIPLE LINES OF VARIOUS SIZES USE BOP EL

SHOW LARGER PIPE DOUBLE LINE

A

A

SECTION A-A

OPEN

GATE

OPEN

GLOBE

ARC OF TRAVEL

PLUG (WRENCH OPERATED)

PLUG (GEAR OPERATED)

MISCELLANEOUS ARRANGEMENTS (continued)

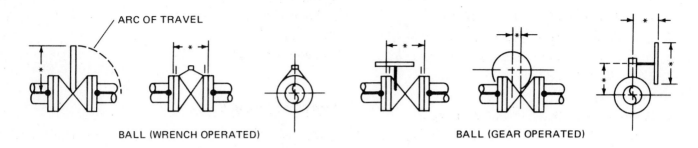

BALL (WRENCH OPERATED)   **BALL (GEAR OPERATED)**

SINGLE LINE VALVES
*DRAW VIEWS OF VALVE PARTS TO SCALE

Fig. 6-45 Quick reference fitting-symbol chart.

EXERCISES

6-1. Define the following terms:

1. Fitting	23. Eccentric reducer
2. Bell-and-spigot	24. Stub in
3. Schedule	25. Reducing elbow
4. Pipe weight	26. Weld cap
5. Pipe bore	27. Header
6. Return bend	28. Branch
7. Elbow	29. Weld symbol
8. Butt weld	30. Edge-of-metal
9. Long radius	preparation
10. Short radius	31. Arrow side
11. Nominal diameter	32. Other side
12. 45° ell	33. Reference line
13. Cutback ell	34. Finish
14. Arc length	35. Flange
15. Center to face	36. Pitch of weld
16. Mitred elbow	37. Code
17. Straight tee	38. Threaded fittings
18. Reducing tee	39. Coupling
19. Lateral	40. Union
20. Reducing Lateral	41. Socket fitting
21. Welding cross	42. Bonney fitting
22. Concentric reducer	

6-2. Match the terms and abbreviations:

1. Ell	a. Butt weld	5. nom. dia.	e. Coupling
2. S.R.	b. Centerline	6. C. to F.	f. Fitting
3. L.R.	c. Concentric reducer	7. C_x	g. Header
4. B.W.	d. Elbow	8. str. tee	h. Reducing tee
		9. red. tee	i. Nominal diameter
		10. conc. red.	j. Long radius
		11. ecc. red.	k. Short radius
		12. hdr.	l. Center to face
		13. ref. line	m. Eccentric reducer
		14. thrd.	n. Straight tee
		15. cplg.	o. Schedule
		16. ftg.	p. Typical
		17. sch.	q. Steel
		18. typ.	r. Threaded
		19. stl.	s. Reference line
		20. C.I.	t. Cast iron
		21. smls.	u. Extra strong
		22. X-strong	v. Seamless
		23. XX-strong	w. Blind flange
		24. std.	x. Double extra strong
		25. S.O.	y. Standard
		26. B.F.	z. Connection
		27. dims.	aa. Dimensions
		28. assy.	ab. Material
		29. conn.	ac. Slip-on
		30. mtl.	ad. Threadolet
		31. M.I.	ae. Weldolet
		32. W.O.L.	af. Assembly
		33. S.O.L.	ag. Malleable iron
		34. T.O.L.	ah. Elbolet
		35. E.O.L.	ai. Sockolet

6-3. Look up the following fitting and flange dimensions in reference charts:

1. Actual O.D. 6″ nom. pipe stainless
2. Wall thickness sch. 30, 8″ pipe stainless
3. Actual O.D. 3″ nom. pipe wrought steel
4. Y 6″, 300# flange
5. No. of bolts 24″, 150# flange
6. A dim. of 4″, 2,000# threaded fitting
7. L dim. of ½″, 3,000# socket-weld fitting
8. U dim. of 2½″, 3,000# socket-weld fitting
9. B dim. of a 6″ butt-weld fitting
10. A dim. of 14″ butt-weld fitting
11. A dim. of 6″, 90° elbow long radius
12. D dim. of a 300# cast-steel gate valve

6-4. Look up socket-weld standard sizes in reference charts.

1. Overall dimension 2″, 200# tee
2. Center to face 4″, 2,000#, 90° elbow
3. M dim. 4″ sch. 40 cross
4. S dim. 2½″, sch. 160 reducer
5. V dim. 3″, 2,000# cap
6. C dim. 3″, 4,000#, 45° Y bend
7. D dim. 2″, 6,000#, 90° elbow
8. N dim. of 2″, 2,000#, 45° Y bend
9. F dim. of 2″, 2,000#, 45° elbow
10. U dim. of 4″ × 2″, 2,000# reducer

6-5. Look up standard sizes of butt-weld fittings in standard reference charts. Determine the following dimensions:

Elbows
1. Center to face of a 12″, 90° L.R. ell
2. Center to face of a 6″, 45° ell
3. Center to face of a 8″ × 6″ red. ell
4. Center to face of a 24″ S.R. 90° ell
5. Center to face of a 10″, 90° L.R. ell

Tees
6. Total length of an 8″ str. tee
7. Center to face of a 10″ str. tee
8. Center to face of the branch of a 12″ × 10″ red. tee

Reducers
9. Face to face of a 14″ × 6″ conc. red.
10. Face to face of an 8″ × 4″ ell red.

Wall Thickness of Fittings
11. 16″ schedule 40
12. 30″ schedule light wall

Actual O.D. of Pipe
13. nom. O.D. 16″
14. nom. O.D. 3″
15. nom. O.D. 10″

DRAWING PROBLEMS

6-6. Convert two-line pipe layouts to one-line schematics (see Fig. 6-46).

6-7. Redraw view given as a schematic and add plan and right side elevation (see Fig. 6-47).

6-8. From plan view given, draw right elevation, left elevation, top, and bottom views (see Fig. 6-48).

6-9. Make layouts for the miter elbows in Figs. 6-49, 6-50, and 6-51 (size to be assigned by instructor).

6-10. Use the data in Fig. 6-52 to make cutback layout and calculations for a 32° elbow L.R. 6″ dia. Lay out a cut for a 90° L.R. elbow, lay out a back to a 40° elbow, and make mathematical calculations of new center-to-face dimension.

Note: A 90° L.R. elbow is shown for illustration purpose only. If angle to be cut were less than 45° it would be cut from a 45° elbow; if more than 45°, it would be cut from a 90° elbow (see Fig. 6-52).

REVIEW QUESTIONS

6-11. Define a pipe fitting.

6-12. List eight types of butt-weld fittings.

6-13. How do short-radius and long-radius 90° elbows differ?

6-14. What is the formula for figuring cutback of elbows?

6-15. Sketch four types of miter elbows.

6-16. What are the two basic types of weld tees?

6-17. What is the main advantage of a reducing elbow?

6-18. What is a blind flange used for?

6-19. What are three basic types of weld processes?

6-20. List five types of joint connections in welding.

6-21. How do socket-weld and butt-weld fittings differ?

6-22. List three types of bonney weld fittings.

6-23. What two basic ways can pipe be drawn?

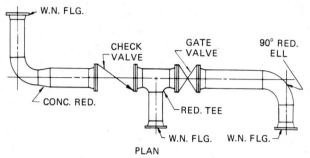

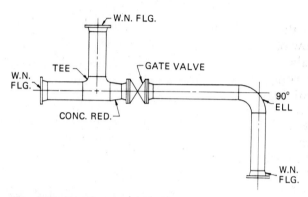

Fig. 6-46 Drawing problem 6-6.

Fig. 6-47 Drawing problem 6-7.

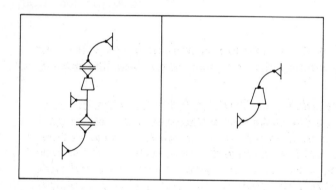

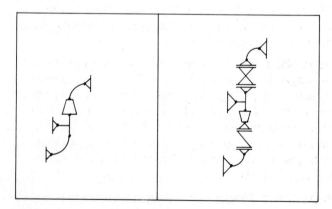

Fig. 6-48 Drawing problem 6-8.

90° MITRE, ONE PIECE

90°

ONE (1) WELD MITRED 90° ELL

(2) 90° MITRED, FOUR-PIECE, 1½ DIA. PIPE

RADIUS

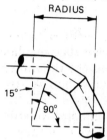

15°

90°

FOUR (4) PIECE MITRED 90° ELL

(3) 45° MITRED ELL TWO PIECE, 1½ DIA. PIPE

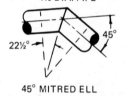

22½°

45°

45° MITRED ELL

Fig. 6-49–51 Drawing problem 6-9.

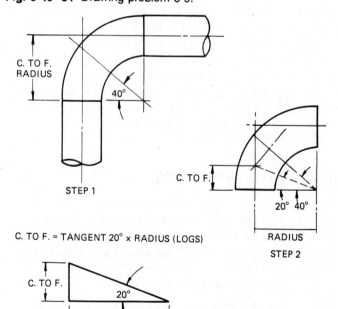

C. TO F. RADIUS

40°

STEP 1

C. TO F.

20° 40°

RADIUS

STEP 2

C. TO F. = TANGENT 20° × RADIUS (LOGS)

C. TO F.

20°

RADIUS

Fig. 6-52 Drawing problem 6-10.

Seven

Valves

This chapter is simply entitled "Valves." However, valves are the heart of the piping system in that they regulate the proper flow of liquids, gases, and combinations thereof through the lines. Valves are precision-made instruments ranging in size from small to very large. There are many types of valves, each having applications for which it is best suited. In this chapter we will discuss some of the most common types of valves, how they operate and what they are used for. After completion of this chapter the student should be able to identify basic types of valves and discuss their operation and application. The student should also be able to use charts and catalogs to find information necessary for properly detailing valves and selecting the correct valve for a specific opera-

DEFINITION OF A VALVE

A *valve* is a piece of equipment which is installed in a piping system for turning on and shutting off or controlling the flow of liquids or gases. There are several types of valves, each with some special design or feature for the intended service. The most commonly used types are the gate valve, globe valve, check valve, plug valve, and control valve. These valves are designed and rated with the same considerations that are given to the design of pipe. Some of the factors taken into account are temperature, pressure, the fluid or gas being handled, erosion and corrosion, and volume of flow.

FUNCTIONS OF VALVES

Each valve has a particular application for which it is best suited. Although valves may appear the same, they are very sophisticated in some cases and represent a great deal of research, engineering, and testing. See Fig. 7-1 for illustration of the five functions of valves listed below:

Starting and Stopping Flow: This is the service for which valves are most generally used. Gate valves are excellently suited for such service. Their seating design permits fluid to move through the open valve in a straight line with minimum restric-

tion of flow and loss of pressure at the valve. The gate principle is not practical for throttling (see Fig. 7-1[1]).

Regulating or Throttling Flow: Regulating or throttling flow is done most efficiently with globe or angle valves. Their seating design causes a change in direction of flow through the valve body, thereby increasing resistance to flow at the valve. Globe and angle valve disk construction permits closer regulation of flow. These valves are seldom used in sizes above 12" due to the difficulty of opening and closing the larger valves against line pressure (see Fig. 7-1[2]).

Preventing Back Flow: Check valves perform the single function of checking or preventing reversal of flow in piping. They come in two basic types, swing check and lift check. Flow keeps these valves open, and gravity and reversal of flow close them automatically. As a general rule, swing checks are used with gate valves and lift checks with globe valves (see Fig. 7-1[3]).

Regulating Pressure: Pressure regulators are used in lines where it is necessary to reduce incoming pressure to the required service pressure. They not only reduce pressure but also maintain it at the point desired. Reasonable fluctuations of inlet pressure to a regulator valve do not affect the outlet pressure for which it is set (see Fig. 7-1[4]).

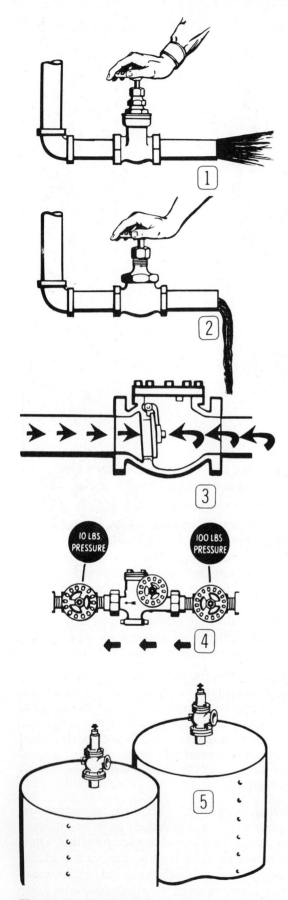

Fig. 7-1 Various functions of valves.

Relieving Pressure: Boilers and other equipment subject to damage from excessive pressures should be equipped with safety valves or relief valves. They usually are spring-loaded valves which open automatically when pressure exceeds the limit for which the valve is set. Safety valves are generally used for steam, air, or other gases. Relief valves are usually used for liquids (Fig. 7-1[5]).

BASIC TYPES OF VALVES

There are dozens of different types of valves, but nine or so are perhaps the most frequently used and could be considered the basic types. Figure 7-2 pictures these basic types, and the following pages of this text give more information regarding them.

SPECIAL TYPES OF VALVES

Each manufacturer builds many valves that are considered special. By *special*, we mean that the valve does a particular job, the design is patented by that manufacturer, and no one else builds one exactly like it. In order to show even a portion of them it would take numerous pages and explanations. Figure 7-3 (p. 76) shows just a few that Crane Company builds. For additional ones see catalogs from other valve manufacturing companies.

SELECTION OF PROPER VALVE AND TYPES OF VALVE MATERIAL

Valve Selection: When in doubt as to which valves and fittings are best suited, do not guess. It is a risky way to equip a pipe system, and it is completely unnecessary. The manufacturer's catalog gives the exact information needed.

First of all, know where the piping materials are going to be used. Will the pressure and temperature installation be high or low? What kind of fluids will be sent through them? Will the conditions of operation be moderate or severe? How much headroom must be allowed for valve stems? What size will the pipelines be? Will the valves have to be dismantled frequently for inspection and servicing? Is the installation to be relatively permanent, or must the piping be broken into frequently? When you can answer these questions, you will know which valves to choose. You will know their operating characteristics, which material you will need, their relative strength in service, what kind of end connections are best suited to the installation; in short, you will know which of the many types and sizes are needed to do the job most efficiently.

Types of Valve Material: It pays to know the range of material from which valves are usually

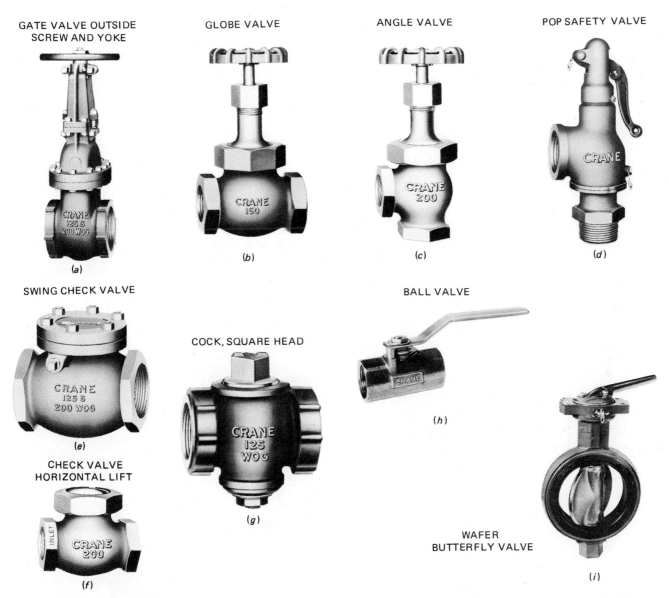

GATE VALVE OUTSIDE
SCREW AND YOKE

(a)

GLOBE VALVE

(b)

ANGLE VALVE

(c)

POP SAFETY VALVE

(d)

SWING CHECK VALVE

(e)

CHECK VALVE
HORIZONTAL LIFT

(f)

COCK, SQUARE HEAD

(g)

BALL VALVE

(h)

WAFER
BUTTERFLY VALVE

(i)

Fig. 7-2 Basic types of valves.

made and to understand the pressure, temperature, and structural limitations of each material. It may be highly unsafe to use materials for services beyond their recommended maximum.

Bronze: *Steam bronze* is an alloy of copper, tin, lead, and zinc. It is widely used in valves and fittings for temperatures up to 450°F [232°C]. *Special bronze* is a high-grade copper-base alloy used in piping equipment for higher pressures and for temperatures up to 550°F.

Iron: Iron is regularly made in three grades: cast iron, ferrosteel, and high-tensile iron. These metals are recommended for temperatures up to 450° [232°C]. Cast iron is commonly used for small valves having light metal sections. Ferrosteel, stronger than cast iron, is used for valves having medium metal thickness. High-tensile iron has even

greater strength, and is used principally for large-size castings.

Malleable Iron: Malleable iron used in valves is characterized by pressure tightness, stiffness, and toughness, the latter being an especially valuable characteristic for piping subjected to stresses and shock.

Steel: Steel is recommended for high pressures and temperatures and for services where working conditions, either internal or external, may be too severe for bronze or iron. Its superior strength and toughness and its resistance to piping strains, vibration, shock, low temperature, and damage by fire afford reliable protection when safety and utility are desired. Many different types of steel—cast, forged, and alloy—are necessary and available because of the widely diversified services steel valves perform.

75

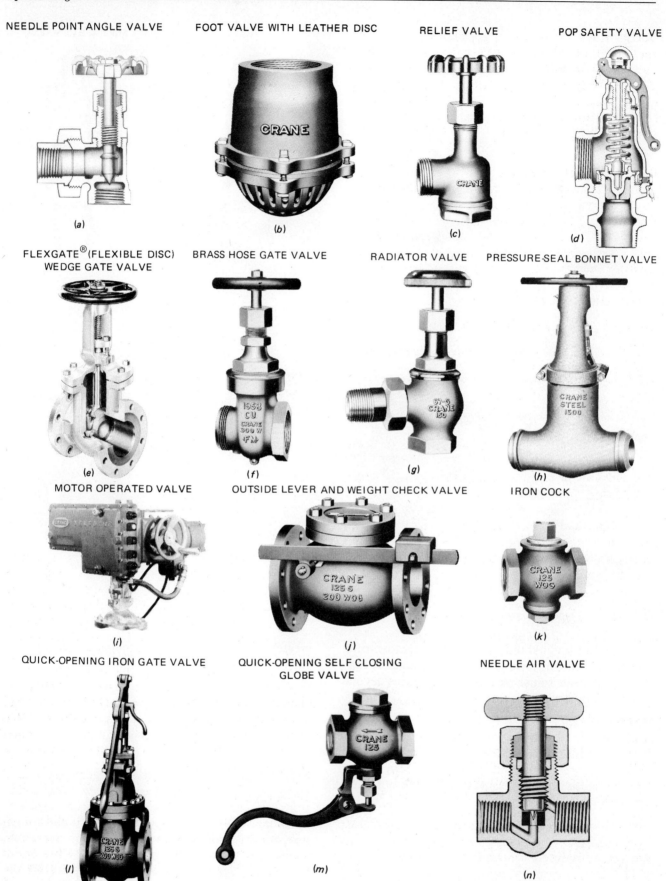

NEEDLE POINT ANGLE VALVE

(a)

FOOT VALVE WITH LEATHER DISC

(b)

RELIEF VALVE

(c)

POP SAFETY VALVE

(d)

FLEXGATE®(FLEXIBLE DISC) WEDGE GATE VALVE

(e)

BRASS HOSE GATE VALVE

(f)

RADIATOR VALVE

(g)

PRESSURE-SEAL BONNET VALVE

(h)

MOTOR OPERATED VALVE

(i)

OUTSIDE LEVER AND WEIGHT CHECK VALVE

(j)

IRON COCK

(k)

QUICK-OPENING IRON GATE VALVE

(l)

QUICK-OPENING SELF CLOSING GLOBE VALVE

(m)

NEEDLE AIR VALVE

(n)

Fig. 7-3 Special types of valves.

Ductile Iron: *Nodular cast iron,* also known as *ductile iron,* is cast iron with the graphite substantially in spherical shape; nodular cast iron is reasonably free of flake graphite. A cast iron having high strength and good ductility results. The corrosion resistance of nodular iron is approximately the same as that of gray iron. Strength is about three times greater.

Stainless Steel: Stainless-steel castings are heat-treated for maximum corrosion resistance, high strength, and good wearing properties. Seating surfaces, stems, and disks of stainless steel are well suited for severe services where foreign materials in the fluids handled could have adverse effects. Stainless steel has excellent resistance to wear, seizure, and oxidation.

GATE VALVE SELECTION

Gate valves are by far the most widely used in industrial piping. That is because most valves are needed as stop valves, that is, to fully shut off or fully turn on flow. This is the only service for which gate valves are recommended (see Fig. 7-4).

Gate valves are inherently suited for wide open service. Flow moves in a straight line and practically without resistance when the operating mechanism (handwheel and stem) fully raises the gate or wedge up into the valve body.

Seating is at right angle to the line of flow and meets it head on. That is one reason why gate valves are impractical for throttling service or for too frequent opening and closing. Repeated movement of the disk near the point of closure under high-velocity flow may create a drag on the seating surfaces and cause galling or scoring on the downstream side. A slightly opened disk may create turbulent flow with vibration and chattering of the disk.

Compared to a globe valve, a gate valve usually requires more turns, consequently more work, to open it fully. Also, the volume of flow through a gate valve is not in direct relation to the number of turns on the handwheel, unlike a globe valve.

Since most gate valves have wedge disks with matching tapered seats, refacing or repairing of the seat is not a simple operation. While not designed for throttling or for too frequent opening and closing, gate valves are generally ideal for services requiring full flow or no flow.

FOUR TYPES OF GATE VALVES

The main variation in the design of a gate valve is in the type of wedge disk used. There are four types:

- Solid wedge disk (Fig. 7-5)

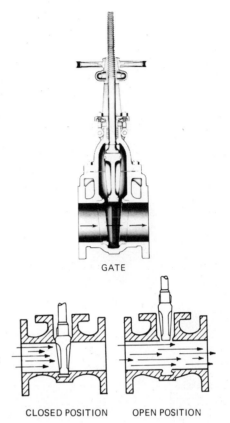

GATE

CLOSED POSITION   OPEN POSITION

Fig. 7-4 Gate valve flow characteristics.

- Double disk (Fig. 7-6)
- Split wedge disk (Fig. 7-7)
- Flexible wedge disk (Fig. 7-8)

Solid Wedge Disk: The most widely used disk in gate valves is the solid wedge-shaped disk with matching tapered body seating surfaces. It is favored for its strong, simple design and single part. It can be installed in any position without danger of jamming from misalignment of parts. It is ideal for steam service, and well suited for water, air, oil, gas, and many other fluids. It is most practical for turbulent flow because there is nothing inside to vibrate and chatter. Refacing of tapered disk surfaces is not easy, but there is little need for it if the valve is used fully open or fully closed. It might be subject to some sticking when subjected to extreme temperature changes if the body contacted more than the disk. For such conditions, the flexible wedge disk is recommended (see Fig. 7-5, p. 78).

Double Disk: The parallel-faced double disk makes closure by descending between matching seats in valve body. On contact with the stop in the bottom of the valve, wedges or spreaders between the disk faces force them squarely against the body seats. The first opening movement releases the disks, and continued operation raises them clear of

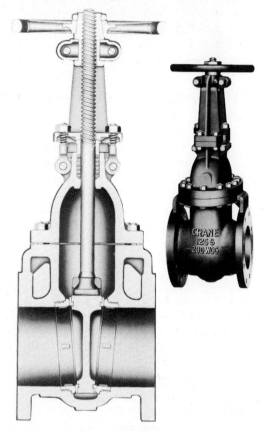

Fig. 7-5 Two views of a solid wedge disk gate valve.

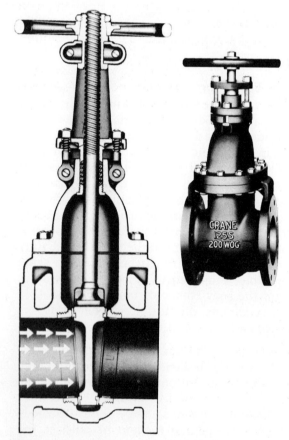

Fig. 7-6 Two views of a double disk gate valve.

the seat openings. This type of disk is widely used on water service, in waterworks and sewage disposal plants, and on oil and gas in cross-country pipelines. It is unsuited for steam because the rapid expansion and high velocity of steam flow tend to vibrate loose internal parts in the disk assembly and hasten their wear. The exposure of a closed valve to an increase in external temperature may cause a dangerous increase in internal pressure if a noncompressible liquid is trapped between the disks. Because disks and body seats are perpendicular and parallel, repairing or refacing to compensate for wear is easier than on a solid wedge disk. The double disk should be installed with the stem above horizontal for best results. Many spreader mechanisms are subject to jamming when installed with the stem below the horizontal line (see Fig. 7-6).

Split Wedge Disk: This is a two-piece wedge disk that seats between matching tapered seats in the body. The spreader device is simple and integral with the disk halves; there are no added parts. When the valve is opened, the first turn releases the disk from the seats (see Fig. 7-7).

Flexible Wedge Disk: This disk was developed especially to overcome sticking in high-temperature service with extreme temperature changes. Solid through the center but not around the outer portion, where it is flexible, its design eliminates sticking and it opens easily under all conditions. It is used in larger size high-pressure, high-temperature valves (see Fig. 7-8). These valves have a sturdy tee-head disk-stem connection. Ample clearance between the head on the stem and the slot in the disk prevents side strains on the stem. The stem is required to move the disk only up and down; it is not forced to serve also as a guide for the disk. The body is made without abrupt changes in the metal section, providing protection against a concentration of stresses produced by pressure, temperature differences and pipeline strains. There are no pockets between the end of the body and the seat to cause turbulence or friction.

TYPES OF STEMS USED ON GATE VALVES

Another variation in valve design is the stem. Some stems rise as the valve is opened; others do not rise. The two types are known as *rising stem* and *non-rising stem*. If a stem has its threads exposed, it is known as an *outside screw and yoke*.

Nonrising Stem with Inside Screw and Yoke: Generally used on gate valves only, this stem does not rise, but merely turns with the handwheel. In turning, the stem threads raise and lower

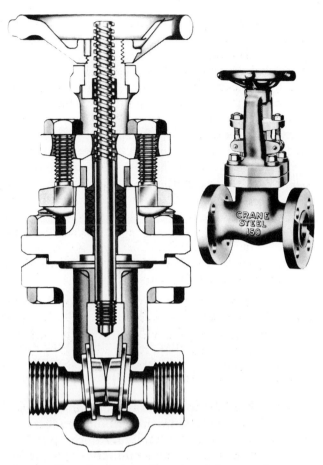

Fig. 7-7 Two views of a split wedge disk gate valve.

Fig. 7-8 A flexible wedge disk gate valve.

the disk. Since the stem only rotates, packing wear is less than if the steam also rose. It is ideal where headroom is limited. Another advantage is that the threads are protected from damage because they are inside.

Rising Stem with Outside Screw and Yoke: Whether the gate valve shown in Fig. 7-5 is open or closed, the stem threads always remain outside the valve body. They are subjected to corrosion and erosion by environmental elements, but not by any elements in the line fluid that might damage stem threads inside the valve body. Being outside, they can be lubricated easily when necessary. Rising-stem valves, by the position of their stems, signal to operators whether they are open or closed. They need adequate headroom.

Rising Stem with Inside Screw: This is the simplest and most common stem construction for gate, globe, and angle valves in the smaller sizes. The stem turns and rises on threads inside the valve. The position of the handwheel indicates the position of the disk, open or closed (see Fig. 7-7).

SEVEN PRESSURE RATINGS FOR FLANGED VALVES

Flanged valve ratings are similar to flange ratings, in that 150-lb raised-face valves have the same flange drilling as 150-lb raised-face flanges; 300-lb raised-face valves have the same flange drilling as 300-lb raised-face flanges; etc.

Valves also have the same variety of facings as flanges—raised face, ring joint, tongue-and-groove, etc. The seven pressures are as follows:

A. 150 lb

B. 300 lb

C. 400 lb

D. 600 lb

E. 900 lb

F. 1,500 lb

G. 2,500 lb

GATE VALVE INSTALLATION INSTRUCTIONS

Installation Note: Valves are usually installed in a vertical position. They can also be installed at any angle determined by the drilling of the companion flange up to 90° from the vertical centerline without resort to rolling the bolt holes to the special angle of the companion flange. It is not recommended to install a valve in a downward position, that is, with the valve in a horizontal position and the handwheel

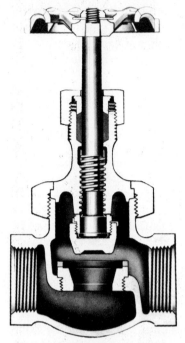

Fig. 7-9 Globe flow characteristics. (This illustration represents a typical Crane globe valve. It is not intended to show all the details and construction of all Crane globe or angle valves).

pointed downward. This downward position makes an obstacle of the handwheel and often hampers operation of the valve.

FUNCTIONS OF GLOBE VALVE

A globe valve derives its name from the globular shape of its body. The valve body must be large enough to allow the full area to open when the valve is in the open position. Globe valves are usually used for regulating flow; however, they can also be used for complete shutoff. The seating in a globe valve is parallel to the flow, whereas the seating in a gate valve is perpendicular (see Fig. 7-9). Flow through a globe valve must make two turns, as shown in Fig. 7-10. A pressure drop results, which is a disadvantange of this design.

ADVANTAGES OF A GLOBE VALVE

The principal advantage of a globe valve over a gate valve is that it is more efficient for the control of flow, and for that reason the globe valve is invariably used for throttling service. Its close regulation is due to the proportional relation of the size of the seat opening to the number of turns of the handwheel, which is the distinctive feature of the plug-type globe valve. An operator can gauge the rate of flow by the number of turns applied to the handwheel.

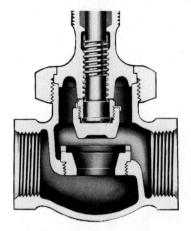

Fig. 7-10 Globe valve plug-type disk.

Time, work, and wear are also saved because of the fewer turns required to operate a globe valve as compared to a gate valve. Whenever wear occurs as the result of frequent or severe operation, the globe presents less of a maintenance problem than a gate valve. Seat and disk in most globe valves can be repaired without removing the valve from the pipeline.

While they are not recommended where resistance to flow and pressure drop would be objectionable, globe valves are generally ideal for throttling: the bypass valve around a control valve, or the steam inlet to a turbine. They are also preferable where frequent operation is necessary.

GLOBE VALVE INSTALLATION

A hand control valve, identified on a flow sheet with the letters H.C. is a globe valve with an indicator attached to it. This indicator tells the operator how high or low the stem is being lowered or raised.

For layout purposes the important dimensions, such as the face-to-face height of stem opened and the diameter of the handwheel, may be found in any valve manufacturer's catalog. The face-to-face dimension has become standardized by the majority of valve manufacturers.

TYPES OF GLOBE VALVE DISKS

Plug-type Disk: A long taper with a corresponding seat, giving a wide area of seating contact, makes the plug-type disk superior to all others for severe throttling service, such as blowoff, soot blower, and boiler feed. Because of its wide seat bearing, most cuts and nicks by dirt, scale, and other foreign matter in the flow are seldom big enough to cause leakage.

The plug disk shape, in the proper combination of metals for service, is most effective in resisting the

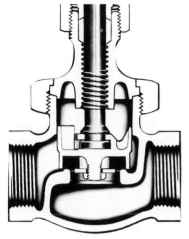

Fig. 7-11 Globe valve ball-type disk.

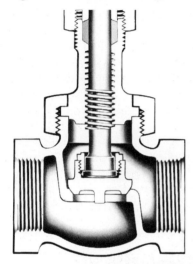

Fig. 7-12 Globe valve composition disk.

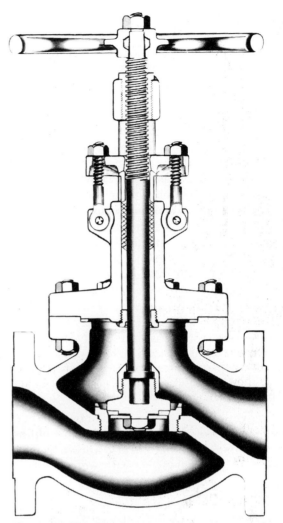

Fig. 7-13 Globe- and angle-valve.

corrosive effects of close throttling. Construction permits replacement of the seat if necessary. Some globe valves have a ball-shaped seat face on the disk. These valves are generally used for steam, water, oil, or oil vapor service.

Conventional Ball-type Disk: This is a design for many not-too-severe services, but not for close throttling. Deposits of particles or foreign matter on the seat make tight closure virtually impossible (see Fig. 7-11).

Composition Disk: This is a useful design in brass and iron valves for adaptability of service and for quick repair. Disks are available in compositions suitable for steam, hot water, cold water, oil, gas, air, gasoline, and many other fluids. Disks can be changed quickly with slip-on disk holders. This design is highly regarded for dependable, tight seating on hard-to-hold fluids such as compressed air. The flat face of the relatively soft disk seats against a raised crown in the body. Small particles of foreign matter are embedded in the disk, preventing seat

damage and leakage. Composition disks are suited for moderate-pressure service with the exception of close regulating and throttling, which can rapidly cut out the disk (see Fig. 7-12).

FUNCTION OF GLOBE-AND-ANGLE VALVE

Globe-and-angle valves are compact and light in weight, and they can be installed in less space than required for conventional valves. Their pressure-seal bonnet assures freedom from bonnet-joint leakage and maintenance. The globe-and-angle valve has a swivel plug-type disk. These valves are used in high-pressure service. The end-to-end dimensions are different from those of flanged valves (see Fig. 7-13).

FUNCTION OF ANGLE VALVE

The internal parts of an angle valve are very similar to those of a globe valve. The angle valve derives its name from the fact that the outlet flow leaves the valve 90° from the inlet.

It is well to remember the angle valve when installing globe valves. If there is a right-angle turn in the line near where a valve is needed, an angle pattern gives important advantages. The angle valve has considerably less turbulence, restriction of flow, and pressure drop than a globe valve, because flow makes one less change of direction. Angle valves also cut down on the piping installation time, labor and materials. An angle valve reduces the number of joints or potential leaks by serving as both a valve and a 90° elbow. The angle valve is available with the same seating variations as the globe valve: plug-type disk, conventional disk, and composition disk. However, many refineries do not encourage the use of angle valves because of standardization programs. An angle valve can be used only at a 90° change of direction, whereas a straight-through globe valve has a more flexible usage, both in location and operability, due to its more advantageous orientation.

Angle valves are used quite extensively in domestic steam-heating systems as shutoff valves at radiators (see Fig. 7-14).

OPERATION OF CHECK VALVE

Check valves have a function all their own: to prevent, or check, a backflow of fluid or gas in a piping system. For this reason they are also known as *non-return valves*. Unlike gate and globe valves, they operate automatically, without a handwheel. They are designed with a swinging flap (the swing check valve) or with a rising disk (the lift check and the stop check valves), which is actuated only by the flow through the valve. The swing check (Fig. 7-15) and lift check (Fig. 7-16) are the basic types most commonly used; the stop check is used only in certain situations. The following is a fuller description of these three types:

Swing Check: The flow through a swing check valve is similar to that through a globe valve in that

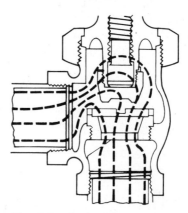

Fig. 7-14 Angle valve.

 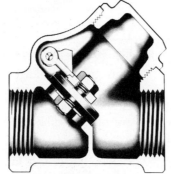 

Fig. 7-15 Swing check valve. (The Y-Pattern swing check should be installed with the direction of flow indicated by the arrow on outside of valve.)

 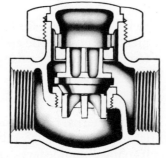 

Fig. 7-16 Lift check valve. (The lift check operates in horizontal lines only.)

it is in a straight line and with no restriction at the seat. This similarity in flow is the reason why a swing check valve is generally used in combination with a gate valve. Should flow reverse, the reverse pressure and the disk's weight close the disk against the seat, and backflow is stopped (see Fig. 7-15).

The swing check valve works automatically, as can readily be seen. The swing check can be installed and operate efficiently in a horizontal or a vertical position for upward flow. But no matter which position is used, this valve will not function unless it is installed with the pressure under the disk.

Lift Check: The flow through a lift check is similar to that through a globe valve in that the path is not in a straight line and therefore the valve offers restriction to the flow. This similarity in flow is the

reason why a lift check valve is generally used with a globe valve. When the flow reverses, the disk falls to its seat and cuts off backflow (see Fig. 7-16).

The lift check valve also operates automatically by line pressure and should be installed with the pressure under the disk.

Stop Check: Another type of swing check valve operates both automatically and manually with the aid of a stem and handwheel. This is called a *stop check valve.* This valve is usually used in stem service (see Figs. 7-17 and 7-18). Stop check valves are recommended for use in a pipeline between each boiler and the main steam header when more than one boiler supplies the steam header. They are also ideally suited for discharge lines from boiler feed pumps or from other high-pressure pumps where a high-quality check valve is needed. Straightway

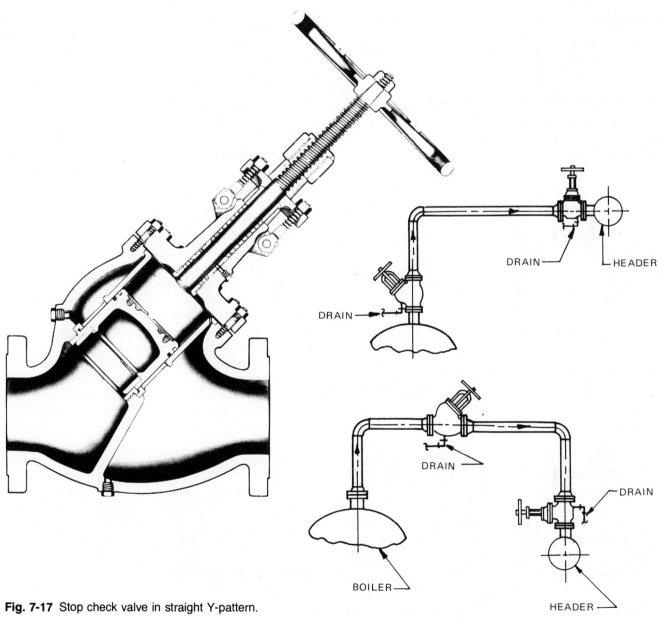

Fig. 7-17 Stop check valve in straight Y-pattern.

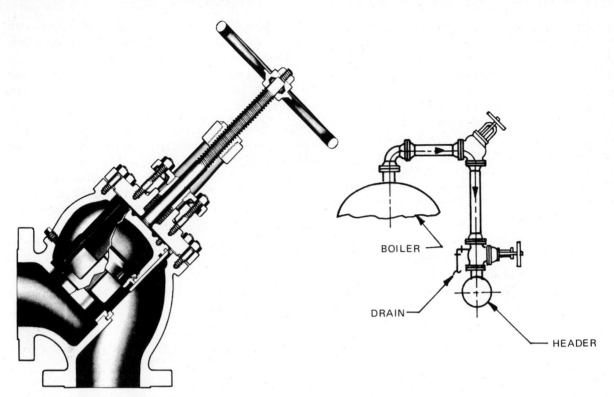

Fig. 7-18 Stop check valve in angle Y-pattern.

valves may be used in horizontal lines or in vertical lines for upward flow. Angle valves are suited for horizontal-downward or upward-horizontal flow.

FUNCTIONS OF STOP CHECK VALVE

Stop check valves are as essential to the safe operation of a boiler plant as safety valves or other safety devices attached to the boiler. The valves are intended to perform four important functions in boiler steam piping:

A. To act as automatic nonreturn valves by preventing a backflow of steam from the main steam header to the boiler to which it is connected, in the event of failure in that boiler.

B. To assist in cutting out a boiler. When the boiler is ceasing to fire, the disk automatically closes and prevents header pressure from entering the boiler.

C. To assist in bringing a boiler into service after a shutdown. This operation requires considerable care when performed manually but is accomplished automatically by a stop check valve without pressure fluctuations or disturbance of the water level.

D. To act as *safety first* valves by preventing backflow of steam from the header into a boiler shut down for repairs or inspection, in the event an attendant accidentally opens the valve.

When more than one boiler is connected to a main steam header, a stop check valve should be installed in the pipeline between each boiler and the header. The valves should always be placed so that the pressure in the boiler is under the disk.

Stop check valves have but one moving part, the disk. Internal parts are easily accessible. When necessary, replacements can be made without removing the body from the line.

These valves are equipped with a handwheel. This permits closing the valves while under pressure or, if already closed, holding the disk in the closed position. There is no mechanical connection between the disk and stem, and when the stem is raised by the handwheel, only the boiler pressure can lift the disk. The assembly for manual operation is the outside screw and yoke type, wherein stem threads are not directly exposed to the high temperature of the steam.

BASIC RELIEF VALVES

The four basic types of relief valves are the *pop safety* relief valve, *relief* valve, *safety* relief valve, and *vacuum*-relief valve. These valves are a necessary part of a refinery for the protection of equipment and piping systems against pressures which exceed the conditions for which they are designed.

The normal operating position of a relief valve is in the closed position with the inlet under the seat.

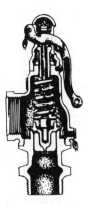

Fig. 7-19 A cross section of a pop safety valve.

The seat is held closed by a spring which is set to restrain a required pressure. If the pressure exceeds the set pressure of the spring, the valve opens and remains open until the pressure subsides.

A relief-valve outlet connection is usually larger than the inlet. The reason for this is that the pressure on the outlet side of the valve is lower than on the inlet side. When the valve opens, the pressure suddenly decreases, causing the volume to increase.

Relief valves require a certain amount of maintenance, and for that reason they should be made accessible from platforms, ladders, etc.

Pop Safety Valves: Pop safety valves are characterized by a pop action and are intended for gas or vapor service. Their design is based on the use of a compressible and therefore expansible fluid which, acting through the seat opening or primary orifice against secondary surfaces in the valve, adds lifting power to the disk, thus causing the valve to suddenly pop open. A large flow area and consequently high capacity are provided. A blowdown regulating ring forms part of the secondary orifice and must be properly adjusted to obtain the desired popping action (see Fig. 7-19).

Relief valves: Relief valves have no pop action. Although suitable for any type of fluid, they are used primarily on liquid service. Their rate of discharge is lower, their action not as positive, and their closing-off pressure not as definite as the blowdown of the pop safety valve. Of the simple poppet type without auxiliary disk-lifting means, they respond to the rise and fall of pressure as rapidly, or gently, as the rate of pressure changes in the system. Liquids such as water and oil are practically noncompressible and do not expand when released to lower pressure regions. The use of a secondary orifice would tend to induce chattering; therefore, these valves do not contain blowdown regulating rings (see Fig. 7-20).

Safety Relief Valve: Safety relief valves are basically like pop safety valves. They have a blowdown

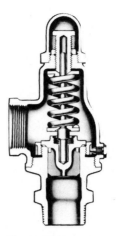

Fig. 7-20 A cross section of a relief valve.

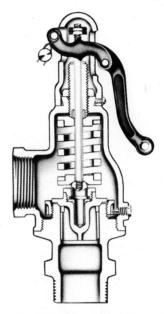

Fig. 7-21 A cross section of a safety relief view.

ring and a huddling chamber and are characterized by a pop action when used on gases or vapors. This design is desirable when the valves are used primarily for pressure relief due to thermal expansion in liquid-laden vessels. As long as the liquid state exists, the flow rate required is low and does not unduly influence performace. When these same vessels tend to generate vapor due to uncontrolled heat input, the low density of the vapor requires a greater discharge; in such cases, the huddling chamber serves to obtain a high disk lift and discharge (see Fig. 7-21).

Vacuum-Relief Valve: The vacuum-relief valve (see Fig. 7-22, p. 86) is designed for use as a vacuum breaker. It is used on low-pressure tanks, steam-jacketed kettles, and similar apparatus where formation of a vacuum could result in damage to the apparatus through the effects of atmospheric pressure.

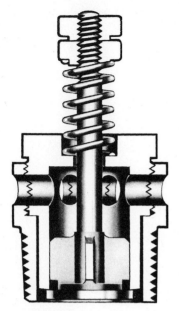

Fig. 7-22 A cross section of a vacuum relief valve.

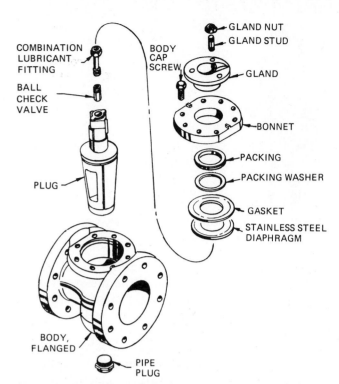

Fig. 7-23 Plug valve assembly.

In service, the disk of the valve is normally held to its seat, both by spring tension and by the pressure in the apparatus to which the valve is attached. When the pressure in the vessel falls below that of the atmosphere, the spring tension is overcome and the disk is pulled from its seat. Air is then admitted through the valve, thus preventing the vacuum from reaching the danger point.

A vacuum-relief has a two-piece body and a seat bushing which removes the seat from the influence of body distortion when the valve is being installed. **Installation Note:** Always install a relief valve in a vertical position (preferably inverted on air and gas service) directly on the apparatus to be protected. Horizontal or angular installation may result in damage to the seating surfaces.

Piping should be carefully planned. Outlet piping, preferably as short as possible with minimum turns, should never be smaller in size than the valve outlet. Inlet piping, if needed, should be full size and not longer than a tee fitting of corresponding size. Stop valves should not be used between the safety valve and the equipment being protected.

Side outlet valves should be used where it is desirable to pipe the discharge away from the immediate vicinity. Discharge piping should be independent of the valve, supported close to the valve, and sloped slightly downward so as to drain away condensate.

FUNCTION OF PLUG VALVES

When a plug valve is in the open position, the flow runs parallel to the pipeline through a hole bored in the plug. To shut off the flow all that is required is a 90° turn, which places the hole in the plug against the body. The plug valve is used similarly to the gate valve in that it is not a valve adaptable for control of flow, as is the globe valve. However, the plug valve is not used as commonly as the gate because it requires more maintenance—specifically, lubrication for good sealing and ease of operation—and because of the possibility of the plug sticking in the open or shut position. Also, it is difficult to see at a glance whether a plug valve is open or shut, unlike a gate valve with a rising steam. A valve wrench is used to open or shut plug valves in size 4″ and below. Valves above 4″ are equipped with a gear and handwheel for easy operation (see Fig. 7-23).

OPERATION OF CONTROL VALVES

Control valves are used to control liquids, gas, and air. The design of the body is similar to that of a globe valve. And as in a globe valve, flow through the seat makes two changes of direction.

Control valves are actuated by the pressure differences in pipelines. This is accomplished with orifice flanges. Pressure indicators register the pressure. Liquid-level controllers on vessels also are used to operate control valves. Air is piped to the liquid-level controller and then to the control valve. The liquid-level controller has a sensor of some type inside the vessel which is attached to an instrument outside the vessel. Air is piped through the instrument to the control valve when the level inside the

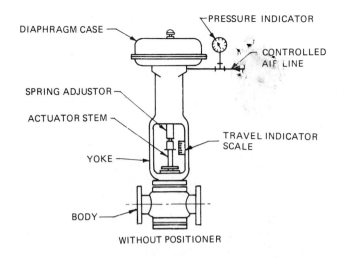

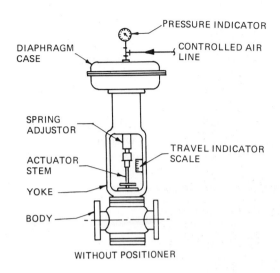

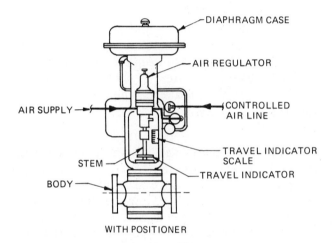

Fig. 7-24 Control valve—air to lift stem.

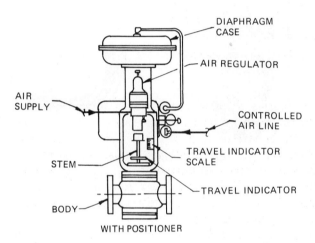

Fig. 7-25 Control valve—air to lower stem.

vessel falls or rises. This opens or closes the control valve. There are two basic types of control valves, the *air to lift stem* (Fig. 7-24) and the *air to lower stem* (Fig. 7-25). Note in Fig. 7-24 that the travel-indicator scale gives quick visual indication of the valve plug position.

The diaphragm case contains a molded fabric diaphragm which by means of air pressure moves the spring to open or close the valve. Diaphragms with air pressure on top are known as *direct-acting actuators.* Increasing air pressure on the diaphragm moves the actuator stem downward. With loss of operating medium, the stem moves to the extreme upward position. Diaphragms with air pressure under the diaphragm are known as *reverse-acting actuators.* Increasing air pressure on the diaphragm moves the actuator stem upward. The stem moves to the extreme downward position upon loss of operating medium.

Control valves also have the same variety of facings as flanges, such as raised face, ring joint, etc.
Installation Note: Control valves are usually installed in a horizontal position with two block valves, one on each side. These block valves are gate valves. There is also a bypass valve, which in most cases is a globe valve. The globe-valve is usually closed unless the control valve needs repair. In that case, the two block valves are closed and the globe valve opened to the desired pressure to keep the pipeline in operation.

BUTTERFLY VALVE

There are three types of butterfly valves: the lug body, the wafer body, and the two-flange body (see Fig. 7-26a, p. 88). The *lug body* has lugs to match the companion flange's bolting pattern. The *wafer body* does not have lugs for bolting; it simply sandwiches between the companion flanges. The two-flange body type has built-in flanges. Gaskets are not required for a butterfly valve since the valve liner itself forms a gasket on both flange faces. Gaskets

87

may be used, however, for protection of the liner where frequent disassembly of the associated piping may be required. Thick, soft gaskets should be avoided.

The word *disk* is marked on the stem below the flats to indicate the position of the disk.

A butterfly valve can be used as a control valve or as a block valve if mounted with a diaphragm. Butterfly valves are used in water service, oil service, etc. They are also used a lot in commercial tanker ships.

When a butterfly valve is in the open position, the flow runs parallel to the pipeline around the disk. To shut off the flow, all that is required is a 90° turn, which places the disk against the valve liner and body.

Installation Note: When the valve opens into a flange with a special bore, etc., ample clearance should be provided so the disk has room to operate properly. (See Fig. 7-26b.)

ONE- AND TWO-LINE VALVE SYMBOLS

Valves are too complex and intricate to draw as they actually appear. Therefore, valve symbols are used to save time and to keep the drawing from being too congested. Figure 7-27 shows the basic valve symbols. Students should become very familiar with them.

Fig. 7-26a Two types of butterfly valves-lug body (left) and water body (right).

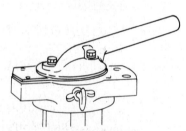

10-POSITION LOCK LEVER
ECONOMICAL, DEPENDABLE. CAN BE LOCKED IN ANY OF 10 CONTROL POSITIONS.

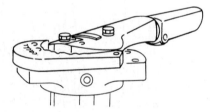

SAFETY TWIST-LOCK 7-POSITION OPERATOR
MAXIMUM SAFETY AGAINST ACCIDENTAL CHANGE OF POSITION. CAN BE PADLOCKED TO PREVENT TAMPERING.

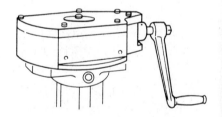

(XP) EXTRA POWER MANUAL SCREW OPERATOR
INFINITE CHOICE OF CONTROL POSITIONS. WEATHERPROOFED. ALSO AVAILABLE WITH HANDWHEEL OR CHAINWHEEL.

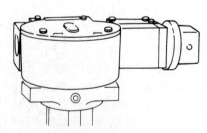

BURIED TYPE RATIO OPERATOR
FOR UNDERGROUND INSTALLATIONS. EXTREMELY DEPENDABLE. COMPLIES WITH AWWA REQUIREMENTS.

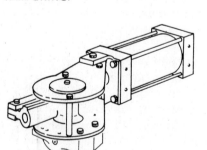

(XP) EXTRA POWER CYLINDER OPERATOR
IDEAL FOR HAZARDOUS OR HARD-TO-REACH LOCATIONS, OR FOR AUTOMATED OPERATION.

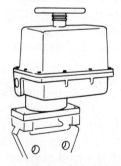

MOTOR OPERATORS
WEATHER-PROOF AND EXPLOSION-PROOF MODELS AVAILABLE FOR ALL REQUIREMENTS.

Fig. 7-26b Butterfly valve—types of operators.

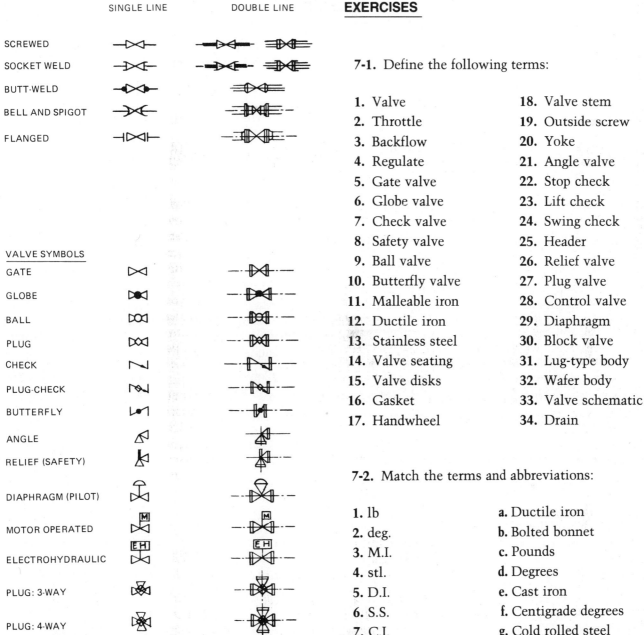

SINGLE LINE   DOUBLE LINE

SCREWED

SOCKET WELD

BUTT-WELD

BELL AND SPIGOT

FLANGED

VALVE SYMBOLS

GATE

GLOBE

BALL

PLUG

CHECK

PLUG-CHECK

BUTTERFLY

ANGLE

RELIEF (SAFETY)

DIAPHRAGM (PILOT)

MOTOR OPERATED

ELECTROHYDRAULIC

PLUG: 3-WAY

PLUG: 4-WAY

Fig. 7-27 Basic valve symbols—single and double line.

CONCLUSION

As was mentioned earlier in this chapter, there are many types of valves and several manufacturers that produce their versions of each. In this chapter we discussed some of the major types of valves and their primary application. The pipe drafting student should become familiar with each of these in addition to others that may be used by the company he or she will be working for. Students should also keep in mind that there are new products appearing on the market daily and it is the drafter's responsibility to keep up with current developments.

EXERCISES

7-1. Define the following terms:

1. Valve	18. Valve stem
2. Throttle	19. Outside screw
3. Backflow	20. Yoke
4. Regulate	21. Angle valve
5. Gate valve	22. Stop check
6. Globe valve	23. Lift check
7. Check valve	24. Swing check
8. Safety valve	25. Header
9. Ball valve	26. Relief valve
10. Butterfly valve	27. Plug valve
11. Malleable iron	28. Control valve
12. Ductile iron	29. Diaphragm
13. Stainless steel	30. Block valve
14. Valve seating	31. Lug-type body
15. Valve disks	32. Wafer body
16. Gasket	33. Valve schematic
17. Handwheel	34. Drain

7-2. Match the terms and abbreviations:

1. lb	**a.** Ductile iron
2. deg.	**b.** Bolted bonnet
3. M.I.	**c.** Pounds
4. stl.	**d.** Degrees
5. D.I.	**e.** Cast iron
6. S.S.	**f.** Centigrade degrees
7. C.I.	**g.** Cold rolled steel
8. B.B.	**h.** Bell and Bell
9. B.M.	**i.** Fahrenheit degrees
10. °F	**j.** Bill of Material
11. °C	**k.** Clean out
12. B. & B.	**l.** Chain-operated
13. B.B.E.	**m.** Manufacturers
14. C.R.S.	**n.** Stainless steel
15. ch. op.	**o.** Malleable iron
16. C.O.	**p.** Steel
17. hdr.	**q.** Bevel both ends
18. mfg.	**r.** Vertical
19. horiz.	**s.** National Pipe Thread

20. vert.	**t.** Horizontal
21. N.P.T.	**u.** Header
22. galv.	**v.** Locked closed
23. Ins.	**w.** galvanized
24. M. & F.	**x.** Stem out
25. L.C.	**y.** Hand control
26. L.O.	**z.** Insulate
27. N.C.	**aa.** Male and female
28. N.O.	**ab.** Locked open
29. S.O.	**ac.** Normally open
30. H.C.	**ad.** Normally closed

7-3. Prepare a letter to be mailed to various manufacturers asking for a complimentary reference material catalog.

7-4. Instructor: Prepare an assignment of questions about and sizes of various parts of valves found in catalogs in the classroom.

7-5. Instructor: Use an actual set of plans for reference and have students prepare a list of all the types of valves used on the job.

7-6. Select ten different valves of various sizes and call three different companies for price quotations.

7-7. Write a comparison of the functions of a gate valve and a globe valve.

7-8. Contact companies or individuals who may have various types of valves in small sizes that they would donate to the school. Clean, paint, and mount the valves on boards and identify them.

7-9. Take an old valve of any size or type and cut away a section so that the inside may be seen. Mount the valve and identify parts.

7-10. Make a list of all the valve suppliers and manufacturers in your area. Include their addresses and telephone numbers.

7-11. Use standard tables to look up the following dimensions:

a. Face-to-face dimension of a 20″, 300# flanged, cast-steel gate valve

b. Center of valve to top of handle open on an 8″, 400# gate valve

c. Handle diameter of an 8″, 150# cast-steel globe valve

d. Center-to-face dimension of a 4″, 900# flanged globe-and-angle valve

e. Face-to-face dimension of a 6″, 300# cast-steel swing check butt-weld valve

f. Centerline to top of a 6″, 300# cast-steel swing check butt-weld valve.

g. N dimension of a 3″, 300# cast-steel swing check screwed valve

h. Open handle height for a 2½″, 900# cast-steel globe valve

7-12. Instructor: Have students disassemble a valve and name all its parts.

DRAWING PROBLEMS

7-13. Make a detail drawing of a gate valve to be assigned by the instructor. Draw interior as a sectional view. (Use valve tables for dimensions and parts list.)

7-14. Convert Fig. 7-28 to a single-line pipe drawing. Represent valves schematically.

7-15. Instructor: Select an area from the set of plans in the Appendix to be redrawn as a single-line pipe drawing with valves shown schematically.

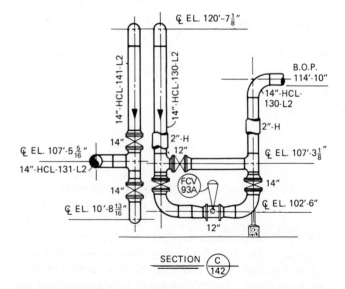

Fig. 7-28 Two-line pipe drawing.

Eight

Piping-Related Math

- Basic arithmetic review
- Angle calculations and *Smoley's* usage
- Calculator orientation
- Metric system conversions and applications

It is assumed at this point that the pipe drafter has had basic mathematics and perhaps other, more advanced courses such as trigonometry and algebra. Math is a very important tool to pipe drafters. Without it, their hands would be tied and their ability to function would be lost. It is the responsibility of the pipe drafter to become acquainted with various types of reference tables, charts, and conversion factors, along with principles of math, to solve required calculations for size, thickness, stress, and other pipe design factors.

BASIC ARITHMETIC REVIEW

Basic arithmetic is essential to pipe calculation, but covering it in minute depth would take very long and perhaps be redundant to the pipe drafter at this point. Basic arithmetic employs the use of

- Whole numbers
- Fractions
- Decimals
- Feet and inches
- The addition, subtraction, multiplication, and division of numbers, fractions, decimals, and feet and inches. If the following explanation is not thorough enough, then a review from a basic textbook in arithmetic would be advisable. Exercises 8-4 to 8-7 at the end of this chapter will help determine how well the student can perform some fundamental operations in basic arithmetic. If the student scores too low, additional review may be needed.

ANGLE CALCULATION

Triangles have been used down through the years as a basis for distance calculations. A triangle forms definite relationships among its sides and angles.

These constant relationships make it possible to use triangles in calculating pipe lengths. In this section we discuss the parts of a triangle, how to label the sides and angles of a triangle, and how to express the relationship of these sides and angles in usable formulas for problem solving. Angles are expressed in degrees, minutes, and seconds.

TRIANGLE NOMENCLATURE

A triangle is made up of six basic parts, three sides and three angles (see Fig. 8-1). There are several types of triangles, but we will deal only with the right triangle (see Fig. 8-2). In a right triangle one of the three angles is a 90° angle.

Label sides and angles with letters (see Fig. 8-3). Use lowercase letters for sides and uppercase letters for angles. The lowercase letter for a side determines the opposite angle's identity.

Side a	vertical leg
Side b	horizontal base
Side c	side opposite 90° angle

Angle A	opposite side a
Angle B	opposite side b
Angle C	opposite side c

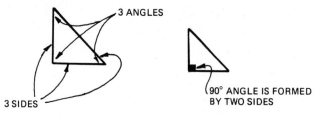

3 ANGLES

3 SIDES

Fig. 8-1 The parts of a triangle.

90° ANGLE IS FORMED BY TWO SIDES

Fig. 8-2 90° right angle.

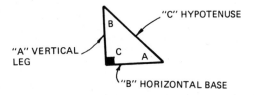

"C" HYPOTENUSE

B

"A" VERTICAL LEG

C

A

"B" HORIZONTAL BASE

Fig. 8-3 Identifying the sides and angles.

Another system of side identification used with basic trig formula calculation has to do with the hypotenuse, adjacent side, and opposite side (see Fig. 8-4a).

After identifying the sides it is possible to solve angle calculation problems by using one of the three formulas listed below. The one selected depends on what sides are given and what angle or angles one needs to know.

$$\text{Sine angle} = \frac{\text{opposite side}}{\text{hypotenuse side}}$$

$$\text{Cosine angle} = \frac{\text{adjacent side}}{\text{hypotenuse side}}$$

$$\text{Tangent angle} = \frac{\text{opposite side}}{\text{adjacent side}}$$

Now try the problems in Fig. 8-4b.

Angles are measured in degrees, minutes, and seconds. Angles can be added, subtracted, multiplied, or divided using degrees, minutes, and seconds. If an angle is converted to a log of that angle, it can be used in a formula with log numbers of the sides. There are 360 degrees in a circle. Each degree has 60 minutes, and each minute has 60 seconds (see samples below). Note how degrees, minutes, and seconds are kept in columns.

Addition

$$\begin{array}{r} 5°\ 8'31'' \\ +\ 8°\ 3'23'' \\ \hline 13°11'54'' \end{array}$$

Subtraction

$$\begin{array}{r} 25°\ 8'13'' \\ -18°\ 3'\ 8'' \\ \hline 7°\ 5'\ 5'' \end{array}$$

Multiplication

$$\begin{array}{r} 18°\ 3'\ 5'' \\ \times\ 4 \\ \hline 72°12'20'' \end{array}$$

Division

$$\begin{array}{r} 4°1'2'' \\ 4\sqrt{16°4'8''} \end{array}$$

Angles can be expressed in degrees, minutes, and seconds if extreme accuracy is required, but they are normally expressed only in degrees and minutes.

Before getting too involved with angle calculations the student should practice the problems in Exercise 8-8 at the end of this chapter.

APPLICATION OF ANGLE CALCULATION

In pipe drafting it is necessary to solve various types of pipe run problems. Pipe does not always run

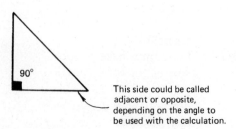

SIDE OPPOSITE 90° (HYPOTENUSE)

90°

90°

This side could be called adjacent or opposite, depending on the angle to be used with the calculation.

Fig. 8-4a Trigonometric side identification.

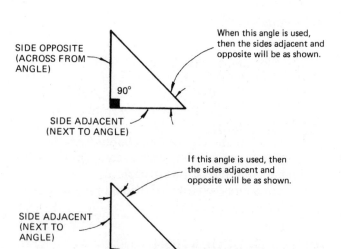

SIDE OPPOSITE (ACROSS FROM ANGLE)

90°

SIDE ADJACENT (NEXT TO ANGLE)

When this angle is used, then the sides adjacent and opposite will be as shown.

If this angle is used, then the sides adjacent and opposite will be as shown.

SIDE ADJACENT (NEXT TO ANGLE)

SIDE OPPOSITE (ACROSS FROM ANGLE)

(a)

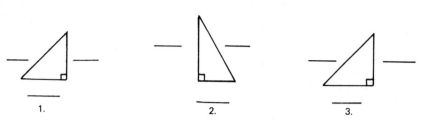

Label the sides of the triangles.
Use letters a, b, and c.

1.   2.   3.

Label the sides of the triangles.
Use hypotenuse, adjacent, and opposite.

4.   5.

Write the formula for solving the
unknown side.

6.

(b)

Fig. 8-4b Trigonometric side identification problems.

up and down or at right angles. Frequently, pipe must run at an angle to miss an obstacle such as a tank or vessel. Occasionally, pipe must run at compound angles such as across a given distance as well as down (see Fig. 8-4c). Or it may go over and up a given distance as in Fig. 8-4d.

SMOLEY'S USAGE

Smoley's Four Combined Tables is a handbook of mathematical tables used to help the drafter perform various math calculations such as squares, bevels, trig functions, and segmental functions. In this unit we cover the following *Smoley's* applications.

- Looking up the square and square root of numbers in feet and inches
- Writing formulas used to determine length of each side of a triangle when other two sides are given
- Using formulas to determine length of unknown sides
- Looking up log of a number and converting a log number to feet and inches

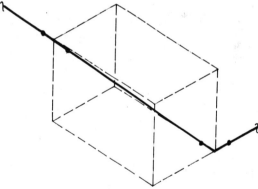

Fig. 8-4c Running pipe at compound angles—"over and down."

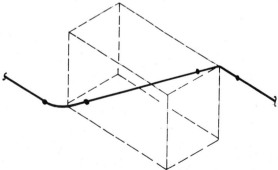

Fig. 8-4d Running pipe at compound angles—"over and up."

68′

Fraction of an inch	0″ Log	0″ Square	1″ Log	1″ Square	2″ Log	2″ Square
0	1.83251	4624.0000	1.83304	4635.3403	1.83357	4646.6944
1/16	1.83254	4624.7084	1.83307	4636.0495	1.83361	4647.4045
1/8	1.83258	4625.4168	1.83311	4636.7588	1.83364	4648.1147
3/16	1.83261	4626.1252	1.83314	4637.4681	1.83367	4648.8249
1/4	1.83264	4626.8338	1.83317	4638.1775	1.83370	4649.5352
5/16	1.83268	4627.5423	1.83321	4638.8870	1.83374	4650.2455
3/8	1.83271	4628.2510	1.83324	4639.5965	1.83377	4650.9558
7/16	1.83274	4628.9597	1.83327	4640.3060	1.83380	4651.6663
1/2	1.83277	4629.6684	1.83331	4641.0156	1.83384	4652.3767
9/16	1.83281	4630.3772	1.83334	4641.7253	1.83387	4653.0873
5/8	1.83284	4631.0860	1.83337	4642.4350	1.83390	4653.7979
11/16	1.83287	4631.7949	1.83341	4643.1448	1.83394	4654.5085
3/4	1.83291	4632.5039	1.83344	4643.8546	1.83397	4655.2192
13/16	1.83294	4633.2129	1.83347	4644.5645	1.83400	4655.9299
7/8	1.83297	4633.9220	1.83351	4645.2744	1.83404	4656.6407
15/16	1.83301	4634.6311	1.83354	4645.9844	1.83407	4657.3516

Fig. 8-5 Portion of *Smoley's* handbook table of squares.

. . Read the square of
68′-1½″ = 4641.0156

. . Read log of 68′-0″ =
1.83251

. . Read square root of
4643.8546 = 68′-1¾″

- Using basic trig function formulas to determine unknown angle when two sides are given
- Using basic trig function formulas to determine unknown side when one side and one angle are given
- Labeling sides of a triangle with rise, run, and slope
- Using tables to determine rise or slope of a triangle with run and pitch given
- Listing five parts of a circular segment

Look up Square and Square Roots of Feet and Inches:

The first section of *Smoley's* is a table of squares and square roots which starts at 0 in and goes up through 299 ft 11⅞ in. By reading feet at the top and inches and fractions at the side or top, it is a simple matter of coordinating down and across for the square or the log (for examples see Fig. 8-5). For the square root of feet and inches, find the number in decimal form and read the table in reverse order.

Unknown-side Calculation:

There are three formulas for determining the length of an unknown side when the other two sides are given:

$$a = \sqrt{c^2 - b^2}$$
$$b = \sqrt{c^2 - a^2}$$
$$c = \sqrt{a^2 + b^2}$$

A right triangle's sides have a constant relationship in that the squares of the two legs added together equal the square of the hypotenuse (side opposite 90° angle; see Fig. 8-6).

If any two of the sides are given, the third side can be found by using one of the formulas shown in Fig. 8-7. Once the correct formula is selected, the mathematical operation can be performed quite easily with the *Smoley's*.

$$c = \sqrt{a^2 + b^2} \qquad c = \sqrt{3^2 + 4^2}$$

Look up the square of 3 and the square of 4:

$$3^2 = 9 \qquad \text{and} \qquad 4^2 = 16$$

Add the two square together:

$$9 + 16 = 25$$

Look in the *Smoley's* for the square root of 25:

$$\sqrt{25} = 5$$

This is a very simple problem and we could have worked it in our heads, but some problems are not quite so easy. Let's take a problem where side c and side a are both given and the subtraction of squares is required (see Fig. 8-8).

Formula $b = \sqrt{c^2 - a^2}$. Find the square of 3 ft 6 in and 2 ft 9 in:

$$(3 \text{ ft } 6 \text{ in})^2 = 12.250 \qquad (2 \text{ ft } 9 \text{ in})^2 = 7.5625$$

Subtract a^2 from c^2

```
 12.2500
-7.5625
 4.6875
```

Find the square root of 4.6875 by looking under the square column for the closest number. The closest number is 4.6832, the square root of which equals 2 ft 1³¹/₃₂ (see Fig. 8-9, p. 95).

Look up Logs of Numbers and Convert Logs to Numbers:

In order to solve some types of triangle problems it is necessary to use the log of an angle with the log of a side length. Finding the log of a side is very easy: Simply look up the feet, inches,

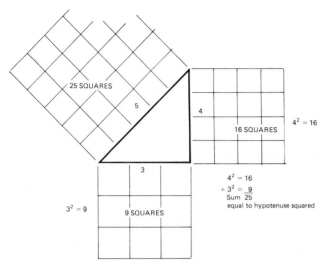

Fig. 8-6 Right angle side relationship.

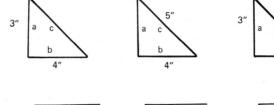

$$c = \sqrt{a^2 + b^2} \qquad a = \sqrt{c^2 - b^2} \qquad b = \sqrt{c^2 - a^2}$$

(a)　　　　(b)　　　　(c)

Fig. 8-7 Basic trigonometric formula for finding sides.

$$b = \sqrt{c^2 - a^2}$$

Fig. 8-8 Using the formula for finding side b.

2′

Fractions of an Inch	0″		1″		2″	
	Log	Square	Log	Square	Log	Square
0	0.30103	4.0000	0.31876	4.3403	0.33579	4.6944
1/32	0.30160	4.0104	0.31930	4.3511	0.33631	4.7057
1/16	0.30216	4.0209	0.31984	4.3620	0.33683	4.7170
3/32	0.30272	4.0313	0.32038	4.3729	0.33736	4.7284
1/8	0.30329	4.0418	0.32092	4.3838	0.33788	4.7397
5/32	0.30385	4.0523	0.32146	4.3947	0.33839	4.7510
3/16	0.30441	4.0627	0.32200	4.4056	0.33891	4.7624
7/32	0.30497	4.0732	0.32254	4.4166	0.33943	4.7738
1/4	0.30553	4.0838	0.32308	4.4275	0.33995	4.7852
9/32	0.30609	4.0943	0.32362	4.4385	0.34046	4.7966
5/16	0.30665	4.1048	0.32415	4.4495	0.34098	4.8080
11/32	0.30721	4.1154	0.32469	4.4605	0.34150	4.8194
3/8	0.30776	4.1260	0.32522	4.4715	0.34201	4.8308
13/32	0.30832	4.1366	0.32576	4.4825	0.34253	4.8423
7/16	0.30888	4.1472	0.32629	4.4935	0.34304	4.8538
15/32	0.30943	4.1578	0.32683	4.5046	0.34355	4.8652
1/2	0.30998	4.1684	0.32736	4.5156	0.34406	4.8767
17/32	0.31054	4.1790	0.32789	4.5267	0.34458	4.8882
9/16	0.31109	4.1897	0.32842	4.5378	0.34509	4.8998
19/32	0.31164	4.2004	0.32895	4.5489	0.34560	4.9113
5/8	0.31219	4.2110	0.32948	4.5600	0.34611	4.9229
21/32	0.31275	4.2217	0.33001	4.5711	0.34662	4.9344
11/16	0.31330	4.2324	0.33054	4.5823	0.34713	4.9460
23/32	0.31385	4.2432	0.33107	4.5934	0.34763	4.9576
3/4	0.31439	4.2539	0.33160	4.6046	0.34814	4.9692
25/32	0.31494	4.2647	0.33212	4.6158	0.34865	4.9808
13/16	0.31549	4.2754	0.33265	4.6270	0.34916	4.9924
27/32	0.31604	4.2862	0.33317	4.6382	0.34966	5.0041
7/8	0.31658	4.2970	0.33370	4.6494	0.35017	5.0157
29/32	0.31713	4.3078	0.33422	4.6607	0.35067	5.0274
15/16	0.31767	4.3186	0.33475	4.6719	0.35118	5.0391
31/32	0.31822	4.3294	0.33527	4.6832	0.35168	5.0508
1	0.31876	4.3403	0.33579	4.6944	0.35218	5.0625

Fig. 8-9 Sample page from *Smoley's* handbook.

and fraction of an inch, such as 68 ft 2½ in (see Fig. 8-10, p. 97). The log of 68 ft 2½ in equals 1.83384.

If a log number such as 1.83406 has been determined and we are trying to find what feet and inches value it has, we would find the closest number under the log column and then read feet, inches, and fraction of inch.

- The feet and inches number for log of 1.83406
- The closest log number is 1.83407, which converts to 68 ft 2¹⁵/₁₆ in.

No matter what the number, it is just that easy.

Basic Trig Functions: The functions are based on the relationship of the hypotenuse, adjacent side, and opposite side (see Fig. 8-11, p. 97).

$$\frac{\text{Side opposite}}{\text{Hypotenuse}} = \text{sine}$$

$$\frac{\text{Side adjacent}}{\text{Hypotenuse}} = \text{cosine}$$

$$\frac{\text{Side opposite}}{\text{Side adjacent}} = \text{tangent}$$

$$\frac{\text{Side adjacent}}{\text{Side opposite}} = \text{cotangent}$$

$$\frac{\text{Hypotenuse}}{\text{Side adjacent}} = \text{secant}$$

$$\frac{\text{Hypotenuse}}{\text{Side opposite}} = \text{cosecant}$$

Although it is good to know all functions, all the problems we deal with can be solved by using only the sine, cosine, and tangent functions.

Determine Unknown Angle When Two Sides Are Given (see Fig. 8-12):

- Label sides opposite, hypotenuse, and adjacent (see Fig. 8-13).
- Based on the two sides that are given, choose the sine, cosine, or tangent formula. The adjacent and opposite sides are given, so use the *tangent* formula.

Tangent is opp./adj., which means if you divide opposite by the adjacent you will get a number that is the tangent angle.

With *Smoley's* you can look up the log of the opposite side, 1 ft 8 in, which is 0.22185; and the log of the adjacent side, 2 ft 2 in, which is 0.33579. Now subtract 0.33579 from 0.22185.

$$\begin{array}{r} 0.22185 \\ -0.33579 \\ \hline 8606 \end{array}$$

We cannot subtract 3 from 1, so by the laws of mathematics we have to make the top number 10 and then go ahead and subtract. The answer is 9.88606:

$$\begin{array}{r} 10.22185 \\ -\ 0.33579 \\ \hline 9.88606 \end{array}$$

Look up this number in the *Smoley's* handbook under the log function index. Look under the tangent column from the top down. The closest number is 9.88603, which gives an angle of 37°34′. If the number could not be found by looking down under the tangent column, you would look up from the bottom above the tangent column.

Problem (see Fig.8-14)

- Label sides (see Fig. 8-15)
- Determine formula; cosine?

$$\text{Angle} = \frac{\text{adjacent}}{\text{hypotenuse}}$$

- Look up log values of adjacent and hypotenuse

$$\frac{\text{Adjacent}}{\text{Hypotenuse}} = \frac{3 \text{ ft } 1 \text{ in} = 0.48902}{4 \text{ ft } 2 \text{ in} \quad 0.61979}$$

Subtract 0.61979 from 0.48902 and again add 10 to the top number.

$$\begin{array}{r} 10.48902 \\ -\ 0.61979 \\ \hline 9.86923 \end{array}$$

Look for angle under cosine 9.86923 (under log function index). The closest value looking under cosine is 9.86924, which is 42°16′. So the answer is 42°16′.

Determine Unknown Side When One Side and One Angle are Given: Using basic trig function formulas to determine unknown side when one side and one angle are given (see Fig. 8-16).

Problem (see Fig. 8-16)

- Label sides (see Fig. 8-17)
- Use side given and side you are trying to find as basis for trig formula. Use hypotenuse-and-opposite formula, which is sine.

- Single angle $= \dfrac{\text{opposite}}{\text{hypotenuse}}$
- Sine 15° $= \dfrac{3 \text{ ft } 8 \text{ in}}{\text{hypotenuse}}$
- Hypotenuse $= \dfrac{3 \text{ ft } 8 \text{ in}}{\text{sine } 15°}$
- Look up log of 3 ft 8 in and log of sine 15°.

log 3 ft 8 in = 0.55247   (make top number 10)

log sine 15° = −9.41300

log sine 15° −9.41300 = 1.14947

68′

Fraction of an Inch	0″		1″		2″	
	Log	Square	Log	Square	Log	Square
0	1.83251	4624.0000	1.83304	4635.3403	1.83357	4646.6944
1/16	1.83254	4624.7084	1.83307	4636.0495	1.83361	4647.4045
1/8	1.83258	4625.4168	1.83311	4636.7588	1.83364	4648.1147
3/16	1.83261	4626.1252	1.83314	4637.4681	1.83367	4648.8249
1/4	1.83264	4626.8338	1.83317	4638.1775	1.83370	4649.5352
5/16	1.83268	4627.5423	1.83321	4638.8870	1.83374	4650.2455
3/8	1.83271	4628.2510	1.83324	4639.5965	1.83377	4650.9558
7/16	1.83274	4628.9597	1.83327	4640.3060	1.83380	4651.6663
1/2	1.83277	4629.6684	1.83331	4641.0156	1.83384	4652.3767
9/16	1.83281	4630.3772	1.83334	4641.7253	1.83387	4653.0873
5/8	1.83284	4631.0860	1.83337	4642.4350	1.83390	4653.7979
11/16	1.83287	4631.7949	1.83341	4643.1448	1.83394	4654.5085
3/4	1.83291	4632.5039	1.83344	4643.8546	1.83397	4655.2192
13/16	1.83294	4633.2129	1.83347	4644.5645	1.83400	4655.9299
7/8	1.83297	4633.9220	1.83351	4645.2744	1.83404	4656.6407
15/16	1.83301	4634.6311	1.83354	4645.9844	1.83407	4657.3516

Fig. 8-10 Finding the log of the number 68′–2¹⁵/₁₆″

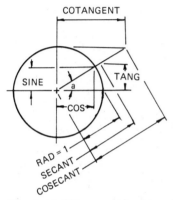

Fig. 8-11 Trigonometric side relationships.

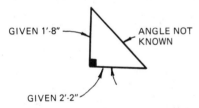

Fig. 8-12 Finding the unknown angle.

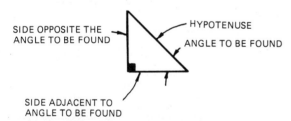

Fig. 8-13 Solving for angle hypoteneuse and given side.

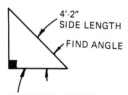

Fig. 8-14 Finding the angle with two given sides.

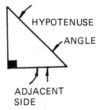

Fig. 8-15 Label sides.

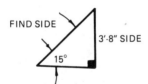

Fig. 8-16 Finding the hypoteneuse with side and angle given.

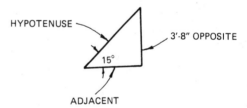

Fig. 8-17 Label sides.

- Find feet and inch number of which 1.14947 is the log (first section of *Smoley's*). The closest number is 1.14947, meaning that 14 ft 1⁹/₃₂ in is the length of the hypotenuse.

In this brief introduction to trig problem-solving we find that there are three basic types of problems that require a solution. All three are based on the right triangle and the relationship of its three sides and three angles. The foregoing trig problems involve the use of functions, of which there are two basic types, the common log functions (L.F.) and natural functions (N.F.). Log functions cannot be used in the same formula with natural functions. A thorough understanding of each is necessary in solving triangular calculation problems. Natural functions are more commonly used since they can be multiplied, subtracted, added, or divided with a calculator. Natural trig functions and a calculator have speeded the process of computation greatly.

Rise, Run, and Slope:

The sides of a right triangle are often expressed as *pitch*, using *rise, run,* and *slope* (see Fig. 8-18). Other terms used to express relationships of the sides to one another are *bevel* and *percent grade.* Each term is used to express a definite ratio among the three sides.

The second section of the *Smoley's* handbook ("Parallel Tables of Slopes and Rises") is a group of tables for bevels from 0 to 12 in. These ratios are expressed as 1:12, 2:12, 4½:12, etc. They mean, for example, that if you go across 12 in and up 4 in, you have a 12:4 bevel relationship. The tables have the calculations already compiled for this relationship for every run of triangle.

To solve a problem with a 12:4 or 4:12 pitch (both mean the same), turn to the 4:12 bevel table in *Smoley's* (see Fig. 8-19). Let's take a run of 10⁷/₁₆ in and follow the steps in Fig. 8-19 for solving the rise and slope.

This principle is the same for all slope, rise, and run problems for whatever run is involved. Practice looking up values for various bevels with different lengths of runs.

Circular Segment:

A circular segment is a portion of a circle. It has five basic parts (see Fig. 8-20, p. 100). Each is indicated with a letter or symbol as shown and can be used in a formula when trying to determine an unknown segment. If any two parts of the five are known, the other three can be found by using the correct formula or table in the segmental functions table in the *Smoley's* handbook. This section is rather lengthy and would need more explanation than we have space for in this text. This area has definite applications and should be covered by a more thorough study of the *Smoley's* handbook. Exercises using the *Smoley's* are included at the end of this chapter (see Exercise 8-9).

CALCULATOR ORIENTATION

Brief History of Calculators: Mathematics itself began long ago in ancient Egypt around the Nile River. Land measurement used forms of geometry, trigonometry, and algebra to establish various types of land markers. As mathematics developed, people became very bored with routine, long-drawn-out forms of addition, subtraction, multiplication, division, and other math functions. At first stones were used to keep track of values in an orderly manner. Later, this method resulted in devices like the *abacus* (which is still being used today).

The evolution of today's calculators was a very slow process at first. Early calculators were mechanical and very slow. The first calculator was invented by Blaise Pascal, a Frenchman who used it for monetary transactions.

With the advent of modern space research the necessity of very small devices and instruments prompted many companies to enter the miniaturization race to see which company could produce the smallest, most versatile, and least expensive hand computer. Hand computers were used at one time by only elite mathematicians, engineers, and technicians. Today the calculator is so small and inexpensive that even a grade school student can afford to purchase one and can learn to operate the basic functions quite easily.

In pipe drafting, the *Smoley's* handbook is gradually being replaced by the calculator, simply because of its speed and accuracy.

Many companies produce calculators, some of which are very efficient. All are similar in many respects, but each has its own individual characteristics. Calculators are getting not only more compact but also less expensive with added capabilities. Careful consideration should be used in selecting the right calculator for the particular job to be performed.

There are several basic types of calculators on the market, each of which has a specific application. The more functions a calculator has, the more it costs and the more difficult it is to operate. Com-

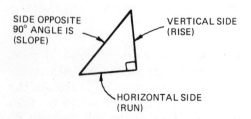

SIDE OPPOSITE 90° ANGLE IS (SLOPE)

VERTICAL SIDE (RISE)

HORIZONTAL SIDE (RUN)

Fig. 8-18 Identifying the rise, run, and slope.

Bevel 4″ TO 12″

Inches	0″ Rise	0″ Slope	1″ Rise	1″ Slope	2″ Rise	2″ Slope	3″ Rise	3″ Slope	4″ Rise	4″ Slope	5″ Rise	5″ Slope
0	0	0	11/32	1 1/16	21/32	2 3/32	1	3 5/32	1 11/32	4 7/32−	1 21/32	5 9/32
1/16	1/32−	1/16	11/32	1 1/8	11/16	2 3/16	1 1/32−	3 7/32	1 11/32	4 9/32	1 11/16	5 11/32−
1/8	1/32	1/8	3/8	1 3/16	23/32−	2 1/4	1 1/32	3 9/32	1 3/8	4 11/32	1 23/32−	5 13/32−
3/16	1/16−	3/16	13/32−	1 1/4	23/32	2 5/16	1 1/16	3 3/8	1 13/32	4 13/32	1 23/32	5 15/32−
1/4	3/32−	1/4	13/32	1 5/16	3/4	2 3/8	1 3/32−	3 7/16	1 13/32	4 15/32	1 3/4	5 17/32−
5/16	3/32	11/32	7/16	1 3/8	25/32−	2 7/16	1 3/32	3 1/2	1 7/16	4 17/32	1 25/32−	5 19/32−
3/8	1/8	13/32−	15/32−	1 7/16	25/32	2 1/2	1 1/8	3 9/16	1 15/32	4 5/8	1 25/32	5 21/32
7/16	5/32−	15/32	15/32	1 1/2	13/16	2 9/16	1 5/32−	3 5/8	1 15/32	4 11/16	1 13/16	5 23/32
1/2	5/32	17/32−	1/2	1 19/32−	27/32−	2 5/8	1 5/32	3 11/16	1 1/2	4 3/4	1 27/32	5 13/16
9/16	3/16	19/32−	17/32−	1 21/32−	27/32	2 11/16	1 3/16	3 3/4	1 17/32−	4 13/16	1 27/32	5 7/8
5/8	7/32−	21/32	17/32	1 23/32−	7/8	2 25/32−	1 7/32−	3 13/16	1 17/32	4 7/8	1 7/8	5 15/16
11/16	7/32	23/32	9/16	1 25/32−	29/32	2 27/32−	1 7/32	3 7/8	1 9/16	4 15/16	1 29/32−	6
3/4	1/4	25/32	19/32−	1 27/32	29/32	2 29/32	1 1/4	3 15/16	1 19/32−	5	1 29/32	6 1/16
13/16	9/32−	27/32	19/32	1 29/32	15/16	2 31/32−	1 9/32−	4 1/32	1 19/32	5 1/16	1 15/16	6 1/8
7/8	9/32	15/16	5/8	1 31/32	31/32−	3 1/32−	1 9/32	4 3/32	1 5/8	5 1/8	1 31/32−	6 3/16
15/16	5/16	1	21/32−	2 1/32	31/32	3 3/32	1 5/16	4 5/32	1 21/32−	5 7/32−	1 31/32	6 1/4

Inches	6″ Rise	6″ Slope	7″ Rise	7″ Slope	8″ Rise	8″ Slope	9″ Rise	9″ Slope	10″ Rise	10″ Slope
0	2	6 5/16	2 11/32−	7 3/8	2 21/32	8 7/16	3	9 1/2	3 11/32−	10 17/32
1/16	2 1/32−	6 3/8	2 11/32	7 7/16	2 11/16	8 1/2	3 1/32−	9 9/16	3 11/32	10 19/32
1/8	2 1/32	6 15/32	2 3/8	7 1/2	2 23/32	8 9/16	3 1/32	9 5/8	3 3/8	10 11/16
3/16	2 1/16	6 17/32	2 13/32−	7 9/16	2 23/32	8 5/8	3 1/16	9 11/16	3 13/32−	10 3/4
1/4	2 3/32−	6 19/32	2 13/32	7 21/32−	2 3/4	8 11/16	3 3/32−	9 3/4	3 13/32	10 13/16
5/16	2 3/32	6 21/32−	2 7/16	7 23/32−	2 25/32−	8 3/4	3 3/32	9 13/16	3 7/16	10 7/8
3/8	2 1/8	6 23/32	2 15/32−	7 25/32−	2 25/32	8 13/16	3 1/8	9 7/8	3 15/32−	10 15/16
7/16	2 3/32	6 25/32	2 15/32	7 27/32−	2 13/16	8 29/32−	3 5/32	9 15/16	3 15/32	11
1/2	2 5/32	6 27/32	2 1/2	7 29/32−	2 27/32−	8 31/32−	3 5/32	10	3 1/2	11 1/16
9/16	2 9/32	6 29/32	2 17/32−	7 31/32	2 27/32	9 1/32−	3 3/32	10 5/32	3 17/32−	11 1/8
5/8	2 7/32	6 31/32	2 17/32	8 1/32	2 7/8	9 3/32−	3 7/32	10 5/32	3 17/32	11 3/16
11/16	2 7/32	7 1/16	2 9/16	8 3/32	2 29/32	9 5/32	3 7/32	10 7/32	3 9/16	11 1/4
3/4	2 1/4	7 1/8	2 19/32−	8 5/32	2 29/32	9 7/32	3 1/4	10 9/32	3 19/32−	11 11/32−
13/16	2 9/32−	7 3/16	2 19/32	8 1/4	2 15/16	9 9/32	3 9/32−	10 11/32	3 19/32	11 13/32−
7/8	2 9/32	7 1/4	2 5/8	8 5/16	2 31/32	9 11/32	3 9/32	10 13/32	3 5/8	11 15/32−
15/16	2 5/16	7 5/16	2 21/32−	8 3/8	2 31/32	9 13/32	3 5/16	10 15/32	3 21/32	11 17/32−

Step 1—Turn to
4-12 Pitch

Step 2—Find Run
of 10 7/16″

Step 3—Read Rise
and Slope

Answer—Rise = 3 15/32″
Slope = 11″

Fig. 8-19 Portion of "to 12" bevel table from *Smoley's* handbook.

mon calculators range from the basic four-function type with storage, square, square root, trig functions, special constants such as π and log, and exponetial values.

Some calculators can be programmed to do basic and complex operations by storing constant values and recalling them to solve a particular set of circumstances. Even more sophisticated calculators use permanent program cards or tapes that are fed into them for a specific sequence of operations.

Drafters work at various levels of responsibility within a company and are required to perform some basic operations such as multiplication, division, addition, subtraction, squares, square root, and trig functions. A design drafter or senior drafter may use many of the programmable calculators as a matter of routine.

Several companies produce calculators, each with a very good instruction manual that explains its individual characteristics. An attempt to explain the use of each calculator in this text would be most exhaustive and in the final analysis would not do justice to the subject. Whatever calculator a student

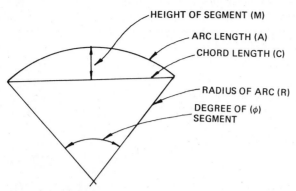

Fig. 8-20 The parts and corresponding symbols of a circular segment.

selects, the main considerations should be quality, cost, efficiency, and versatility.

All calculators require that the individual practice and become very familiar with each key and operation the calculator is capable of performing. The pipe drafter specifically needs to use trig functions, square root, log function, and storage, and perhaps other special keys. Do not spend an excessive amount of money for a calculator. They are a very competitive item and the cost is constantly going down, which is great for the consumer.

METRIC SYSTEM: CONVERSIONS AND APPLICATIONS

The following is a statement made by John Quincy Adams in a report to Congress in 1821. It is a very interesting analysis of the importance of a system of measurement.

> Weights and measures may be ranked among the necessaries of life to every individual of human society. They enter into the economical arrangements and daily concerns of every family. They are necessary to every occupation of human industry; to the distribution and security of every species of property; to every transaction of trade and commerce; to the labors of the husbandman; to the ingenuity of the artificer; to the studies of the philosopher; to the researches of the antiquarian; to the navigation of the mariner and the marches of the soldier; to all the exchanges of peace, and all the operations of war. The knowledge of them, as in established use, is among the first elements of education and is often learned by those who learn nothing else, not even to read and write. This knowledge is riveted in the memory by the habitual application of it to the employments of men throughout life.

A Brief History: People always used weights and measures. Primitive societies, for example, used measurements for their daily tasks: in construction, in bartering, and in clothing.

As societies developed, weights and measures became more complex. With the science of mathematics and the invention of the numbering system, a whole new system of weights and measures evolved. The new system was better suited to trade and commerce, taxation, land division, and science. More sophisticated uses were necessary. Not only did people have to weigh and measure more complex things, but they also had to do it accurately time after time and in different places. It's not surprising that different systems for the same purpose developed. Because there was such little international communication, these different systems became established in different parts of the world. There were even different systems on the same continent.

American Customary System of Measurement: The measurement system that is used most often in the United States today hasn't changed much from what was used by the colonists from England. Their measures originated from a variety of cultures—Babylonian, Egyptian, Roman, Anglo-Saxon, and Norman-French. What was originally digit, palm, span, and cubit units is now inch, foot, and yard. We still don't understand how this transformation occurred.

From Rome came the contribution of the use of the number 12 as a base system. The 12 divisions of the Roman *pes*, or foot, were called *unciae*. This explains why today's foot is divided into 12 inches. Other measures like *inch* and *ounce* are also derived from their Latin meanings.

We can trace the origin of the yard as a measure of length to the early Saxon kings. The Saxon kings wore a girdle or sash around their waists. These were all the same length. And they could be easily removed and used as a convenient measuring device. So, the word *yard* is derived from the Saxon word *gird*—the circumference of a person's waist.

The standardization of these units and their combinations into a related system of weights and measures occurred in various and, at times, fascinating ways. According to tradition, King Henry I decreed that the yard should be the distance from the tip of his nose to the end of his thumb. Similarly, the length of a furlong (or *furrow-long*) was established by early Tudor rulers as 220 yards. And in the sixteenth century, Queen Elizabeth I declared that the traditional Roman mile equaling 5,000 would be replaced by one 5,280 long. This made the mile exactly 8 furlongs. Its point was to provide a convenient relationship between the two earlier ill-related measures.

By the eighteenth century, England had achieved a greater degree of standardization than the rest of Europe. Their units were well suited to business and trade since they had been developed and refined to meet commercial needs.

The Metric System: There are many reasons for changing to the metric system. The United States has had some problems in dealing with the idea of change. It is primarily needed to counteract America's declining position in world trade. And all other major nations already have changed.

There is increased metric use in manufacturing, where the potential exists for the greatest gains in the conversion. There will be principally three kinds of gains:

1. There is a potential increase in exports of products that are manufactured to metric standards. Metric countries give preference to metric design products.

2. There is a savings potential. A common design used for products made by United States companies here and abroad results in a substantial savings.

3. The change to metric designs will reduce the excessive varieties and sizes of products currently made.

Although the United States is already in the process of converting, we are considerably behind all other industrialized nations. It is important to realize that it is industry, not government, that is initiating increased metric use.

In the changeover to metrics, four basic principles are presently being followed:

A. The rule of reason. Changes to metric are made where it is advantageous. Changes are not made just for the sake of change.

B. There are no subsidies.

C. It is a voluntary changeover.

D. There is non-governmental initiative.

Congress has established a national policy to create a mechanism for planning and coordinating the nation-wide conversion. There is no need for the government to do any more than these objectives.

The SI System: The need for a single, world-wide, coordinated measurement system was recognized over 300 years ago. In 1799 the French first adopted the prototype of what was to become the SI metric system. Although the metric system was not initially greeted with enthusiasm, other nations steadily adopted it. It's not surprising that its rapid spread coincided with the age of rapid technical development.

In 1960 the General Convention adopted an extensive revision and simplification of the old metric

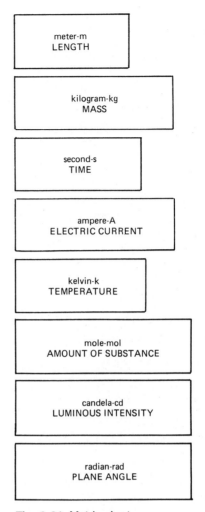

Fig. 8-21 Metric chart.

system, under the name *Le Système international d'unités* (International System of Units), with the abbreviation SI.

In the United States, not much resulted until 1971, when the Department of Commerce recommended a 10-year changeover to the metric system. Many companies have already changed or are in the process of changing even without legislation. It is expected that we will save millions of dollars yearly in increased foreign trade.

The SI Metric System: The SI metric system is based on the base-10 system, which eliminates the need for fractions. Conversions are in multiples of 10, and base units for quantities of length, weight, etc., are easily coordinated and integrated in formulas (see Fig. 8-21). The following seven units of measure are the most commonly used:

A. Meter (length)

B. Kilogram (mass)

C. Second (time)

D. Ampere (electrical current)

SI Metric Unit Prefixes

Value	Multiples and Submultiples	Prefixes
1 000 000 000 000	10^{12}	tera
1 000 000 000	10^{9}	giga
1 000 000	10^{6}	mega
1 000	10^{3}	kilo
1 00	10^{2}	hecto
10	10	deka
0.1	10^{-1}	deci
0.01	10^{-2}	centi
0.001	10^{-3}	milli
0.000 001	10^{-6}	micro
0.000 000 001	10^{-9}	nano
0.000 000 000 001	10^{-12}	pico

Fig. 8-22 Multiples and submultiples of various values.

Linear Measurement

One thousand meters	(10^{3} meters) is a kilometer (km)
One hundred meters	(10^{2} meters) is a hectometer (hm)
Ten meters	(10^{1} meters) is a dekameter (dam)
A meter	(10^{0} meter) is a meter (m)
one tenth of a meter	(10^{-1} meter) is a decimeter (dm)
on hundredth of a meter	(10^{-2} meter) is a centimeter (cm)
one thousandth of a meter	(10^{-3} meters) is a millimeter (mm)

Fig. 8-23 Multiples and submultiples of linear measurements.

E. Celsius (temperature)

F. Mole (molecular weight)

G. Candela (luminous intensity)

Various amounts of each can be expressed by multiples or submultiples of the basic unit, such as the meter. They can be expressed in powers of ten to indicate multiples or submultiples of quantities (see Fig. 8-22).

A person working with metric quantities and prefixes soon determines those that are most common to his or her particular application and learns to increase or decrease a quantity by simply moving the decimal place to the right or left or multiplying as a simple decimal problem. The old coversion problems of one-sixteenth of a yard, gallon, dozen, etc., are eliminated or greatly simplified.

Length Measurement in Metric (Meters): In pipe drafting, measurements of length which are given in feet and inches now will be expressed metrically in multiples or submultiples of meters (see Fig. 8-23).

A meter, also spelled *metre,* is defined in Figs. 8-24, 8-25, and 8-26.

Once we have changed completely to the metric system, the problem of relating the English system to the SI metric system will not occur as frequently as it does now. To help form a mental picture of the relationship of metric values to those in the English system, refer to Figs. 8-27*a* and *b.* Awareness programs to help people "think metric" are essential to a smooth changeover in the system.

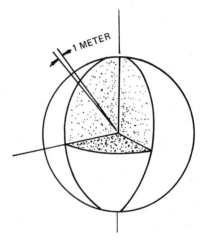

Fig. 8-24 The meter as derived from its relationship to the Earth. The meter is equal to $\dfrac{1}{10,000,000}$ of the distance from the north pole to the equator.

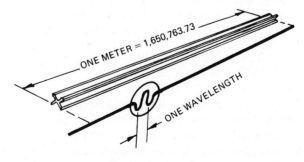

Fig. 8-25 A scientific definition defines the meter as equal to 1,650,763.73 wave lengths of orange-red light given off by Krypton 86.

1 METER

1 YARD

METER: A LITTLE LONGER THAN A YARD
(ABOUT 1.1 YARDS).

1000 millimeters = 1 meter
100 centimeters = 1 meter
1000 meters = 1 kilometer

Fig. 8-26 Comparing a meter (m) to a yard.

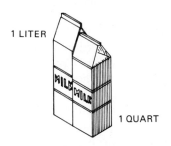

1 LITER

1 QUART

Liter: A little longer than a quart
(about 1.06 quarts).

Fig. 8-27a Comparing a liter (l) to a quart.

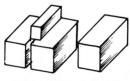

1 KILOGRAM = 1 POUND

Kilogram: 1000 grams = a little more than 2 pounds
(about 2.2 pounds)

Gram: About the weight of a paper clip

COMMON PREFIXES
(to be used with basic units)

Milli = one-thousandth (0.001)
Centi = one-hundredth (0.01)
Kilo = one-thousand times (1000)

Fig. 8-27b A kilogram (kg) as compared to a pound.

In using the metric system to draw pipe, four of the seven units of measure should be emphasized: the kilometer (Fig. 8-28), the gram (Fig. 8-29), the liter (Fig. 8-30), and degrees Celsius (Fig. 8-31, p. 104). Each unit is based on the decimal system of base 10. The figures indicated for each should help clarify their relative value compared to English measurements.

Reading the Metric Scale: The Pipe drafter should be able to read measurements in metric or the customary English dimension system. The basic principle of reading the metric scale at any ratio is primarily the same. Units are in multiples of ten. Whether the ratio is 1:1 or 1:5 or 1:10 (see Fig. 8-32, p. 104), the reading procedure is the same except

KILOMETER

MILE   1/2 MILE   5/8 MILE

Fig. 8-28 A kilometer is used to measure long distances. 1000 meters equal one kilometer; it is equal to about ⅝ of a mile.

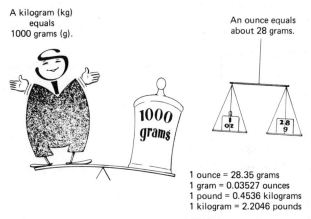

A kilogram (kg)
equals
1000 grams (g).

An ounce equals
about 28 grams.

1 ounce = 28.35 grams
1 gram = 0.03527 ounces
1 pound = 0.4536 kilograms
1 kilogram = 2.2046 pounds

Fig. 8-29 Gram and ounce comparisons.

1 Liter   1 Quart

Fig. 8-30 The liter is about 6 pecent larger than a quart.

that each scale gives a different value to each division.

The most common scale ratios of the SI. metric system are as follows: 1:1, 1:2, 1:5, 1:10, 1:20, 1:50, 1:2,500. The first scale that should be studied is the 1:1, which is full scale or actual size. The scale is graduated in tenths and each mark represents one millimeter (see Fig. 8-33, p. 104).

The SI metric scale, the English scale, and decimals are compared in Fig. 8-34, p. 104.

Conversion Factors: Metric values at times must be converted to English values and require the

103

Celsius—
the Unit of Temperature

In the metric system, scientists use the official base unit of kelvin (k) for temperature. However, the Celsius scale is used for everyday needs. The Celsius scale is derived from the kelvin scale, and they are related by the exact difference of 273.15 degrees.

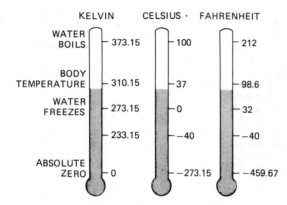

Temp. °K = Temperature °C + 273.15
Temp. °C = $\dfrac{\text{Temp. F-32}}{1.8}$
Temp. °F = 1.8 (Temp. °C) + 32

Fig. 8-31 Temperature comparisons.

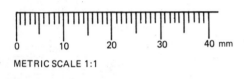

METRIC SCALE 1:1

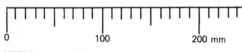

METRIC SCALE 1:5

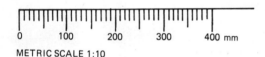

METRIC SCALE 1:10

Fig. 8-32 Metric scale ratios.

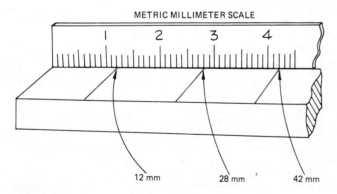

Fig. 8-33 Reading the millimeter (mm) scale.

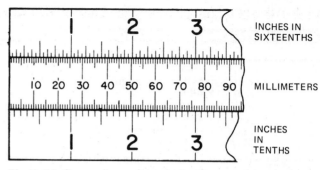

Fig. 8-34 Comparing metrics to fractions and decimals.

use of a conversion table, or multiplication by a constant factor to derive the needed value. Conversion within the metric system itself can be accomplished by moving the decimal place to the left or right.

There are two ways to convert to or from the metric system. First, use metric conversion constants and perform basic math calculation (see Fig. 8-35). Second, use a table with numerous conversions already calculated (see Fig. 8-36 and on page 106, Figs. 8-37 and 8-38).

Complete conversion to metric involves an avalanche of problems and must be planned carefully to insure a smooth transition. Many companies that supply pipe, valves, and equipment have already converted and provide charts and reference material for the drafter to use on jobs that are using the metric system. Practice conversions and calculations by working the exercises at the end of this chapter (see Figs. 8-39 and 8-40, p. 107).

Approximate Conversions from Customary to Metric and Vice Versa

	When You Know:	You Can Find:	If You Multiply By:
LENGTH	inches	millimeters	25.4*
	feet	centimeters	30.48*
	yards	meters	0.9
	miles	kilometers	1.6
	millimeters	inches	0.04
	centimeters	inches	0.04
	meters	yards	1.1
	kilometers	miles	0.6
AREA	square inches	square centimeters	6.5
	square feet	square meters	0.09
	square yards	square meters	0.8
	square miles	square kilometers	2.6
	acres	square hectometers (hectares)	0.4
	square centimeters	square inches	0.16
	square meters	square yards	1.2
	square kilometers	square miles	0.4
	square hectometers (hectares)	acres	2.5
MASS	ounces	grams	28
	pounds	kilograms	0.45
	short tons	megagrams (metric tons)	0.9
	grams	ounces	0.035
	kilograms	pounds	2.2
	megagrams (metric tons)	short tons	1.1
LIQUID VOLUME	ounces	milliliters	30
	pints	liters	0.47
	quarts	liters	0.95
	gallons	liters	3.8
	milliliters	ounces	0.034
	liters	pints	2.1
	liters	quarts	1.06
	liters	gallons	0.26
TEMPER-ATURE	degrees Fahrenheit	degrees Celsius	5/9 (after subtracting 32)
	degrees Celsius	degrees Fahrenheit	9/5 (then add 32)

Fig. 8-35 Converting constants.

Fractions of an Inch to Millimeters

in	mm		in	mm		in	mm		in	mm	
1/64	0.016	0.397	17/64	0.266	6.747	33/64	0.516	13.097	49/64	0.766	19.447
1/32	0.031	0.794	9/32	0.281	7.144	17/32	0.531	13.494	25/32	0.781	19.844
3/64	0.047	1.191	19/64	0.297	7.541	35/64	0.547	13.891	51/64	0.797	20.241
1/16	0.062	1.588	5/16	0.312	7.938	9/16	0.562	14.288	13/16	0.812	20.638
5/64	0.078	1.984	21/64	0.328	8.334	37/64	0.578	14.684	53/64	0.828	21.034
3/32	0.094	2.381	11/32	0.344	8.731	19/32	0.594	15.081	27/32	0.844	21.431
7/64	0.109	2.788	23/64	0.359	9.128	39/64	0.609	15.478	55/64	0.859	21.828
1/8	0.125	3.175	3/8	0.375	9.525	5/8	0.625	15.875	7/8	0.875	22.225
9/64	0.141	3.572	25/64	0.391	9.922	41/64	0.641	16.272	57/64	0.891	22.622
5/32	0.156	3.969	13/32	0.406	10.319	21/32	0.656	16.669	29/32	0.906	23.019
11/64	0.172	4.366	27/64	0.422	10.716	43/64	0.672	17.066	59/64	0.922	23.416
3/16	0.188	4.762	7/16	0.438	11.112	11/16	0.688	17.462	15/16	0.938	23.812
13/64	0.203	5.159	29/64	0.453	11.509	45/64	0.703	17.859	61/64	0.953	24.209
7/32	0.219	5.556	15/32	0.469	11.906	23/32	0.719	18.256	31/32	0.969	24.606
15/64	0.234	5.953	31/64	0.484	12.303	47/64	0.734	18.653	63/64	0.984	25.003
1/4	0.250	6.350	1/2	0.500	12.700	3/4	0.750	19.050	1	1.000	25.400

Fig. 8-36 Converting fractions of an inch to millimeters.

Millimeters to Inches

mm	in	mm	in	mm	in	mm	in
1	0.039	26	1.024	51	2.008	76	2.992
2	0.079	27	1.063	52	2.047	77	3.031
3	0.118	28	1.102	53	2.087	78	3.071
4	0.157	29	1.142	54	2.126	79	3.110
5	0.197	30	1.181	55	2.165	80	3.150
6	0.236	31	1.220	56	2.205	81	3.189
7	0.276	32	1.260	57	2.244	82	3.228
8	0.315	33	1.299	58	2.283	83	3.268
9	0.354	34	1.338	59	2.323	84	3.307
10	0.394	35	1.378	60	2.362	85	3.346
11	0.433	36	1.417	61	2.402	86	3.386
12	0.472	37	1.457	62	2.441	87	3.425
13	0.512	38	1.496	63	2.480	88	3.464
14	0.551	39	1.535	64	2.520	89	3.504
15	0.590	40	1.575	65	2.559	90	3.543
16	0.630	41	1.614	66	2.598	91	3.583
17	0.669	42	1.654	67	2.638	92	3.622
18	0.709	43	1.693	68	2.677	93	3.661
19	0.748	44	1.732	69	2.716	94	3.701
20	0.787	45	1.772	70	2.756	95	3.740
21	0.827	46	1.811	71	2.795	96	3.780
22	0.866	47	1.850	72	2.835	97	3.819
23	0.906	48	1.890	73	2.874	98	3.858
24	0.945	49	1.929	74	2.913	99	3.898
25	0.984	50	1.968	75	2.953	100	3.937

Fig. 8-37 Converting millimeters to inches.

Inches to Millimeters

in	mm	in	mm	in	mm	in	mm
1	25.4	26	660.4	51	1295.4	76	1930.4
2	50.8	27	685.8	52	1320.8	77	1955.8
3	76.2	28	711.2	53	1346.2	78	1981.2
4	101.6	29	736.6	54	1371.6	79	2006.6
5	127.0	30	762.0	55	1397.0	80	2032.0
6	152.4	31	787.4	56	1422.4	81	2057.4
7	177.8	32	812.8	57	1447.8	82	2082.8
8	203.2	33	838.2	58	1473.2	83	2108.2
9	228.6	34	863.6	59	1498.6	84	2133.6
10	254.0	35	889.0	60	1524.0	85	2159.0
11	279.4	36	914.4	61	1549.4	86	2184.4
12	304.8	37	939.8	62	1574.8	87	2209.8
13	330.2	38	965.2	63	1600.2	88	2235.2
14	355.6	39	990.6	64	1625.6	89	2260.6
15	381.0	40	1016.0	65	1651.0	90	2286.0
16	406.4	41	1041.4	66	1676.4	91	2311.4
17	431.8	42	1066.8	67	1701.8	92	2336.8
18	457.2	43	1092.2	68	1727.2	93	2362.2
19	482.6	44	1117.6	69	1752.6	94	2387.6
20	508.0	45	1143.0	70	1778.0	95	2413.0
21	533.4	46	1168.4	71	1803.4	96	2438.4
22	558.8	47	1193.8	72	1828.8	97	2463.8
23	584.2	48	1219.2	73	1854.2	98	2489.2
24	609.6	49	1244.6	74	1879.6	99	2514.6
25	635.0	50	1270.0	75	1905.0	100	2540.0

Fig. 8-38 Converting inches to millimeters.

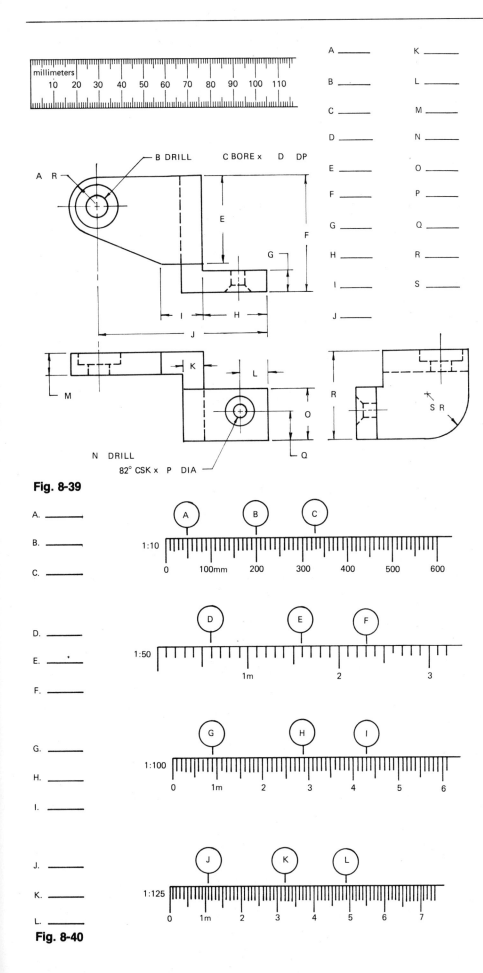

Fig. 8-39

A. _____
B. _____
C. _____

D. _____
E. _____
F. _____

G. _____
H. _____
I. _____

J. _____
K. _____
L. _____

Fig. 8-40

EXERCISES

8-1. Define the following terms in your own words.

1. Triangle
2. Whole number
3. Mixed number
4. Fraction
5. Decimal
6. Degree
7. Minute
8. Second
9. Right triangle
10. Hypotenuse
11. Adjacent
12. Opposite
13. Square of a number
14. Square root of a number
15. *Smoley's* handbook
16. Basic trig function
17. Common log function
18. Natural log function
19. Rise
20. Run
21. Slope
22. Pitch
23. Bevel
24. Circle segment
25. Metric system
26. Calculator
27. Linear measurement
28. Common denominator
29. Numerator
30. Reduce to lowest term
31. Decimal equivalent
32. Fractional equivalent
33. Sine
34. Chord
35. Tangent
36. Millimeter
37. Centimeter
38. Decimeter
39. Dekameter
40. Kilometer
41. Power of a number
42. Cosine

8-2. Write the word that describes the symbols listed.

1. $'$
2. $''$
3. $°$
4. $=$
5. $\cap$
6. $\perp$
7. $\parallel$
8. $+$
9. $-$
10. $\times$
11. $\div$
12. $\emptyset$

8-3. Write the word that the abbreviation stands for.

1. hyp.
2. adj.
3. opp.
4. L.F.
5. N.F.
6. trig.
7. log
8. bev.
9. km
10. hm
11. dam
12. m
13. dm
14. cm
15. mm
16. natl. log

8-4. Basic Math Problems: Whole Numbers

Add and check:

1. $7 + 3 + 9 + 4 + 8 + 16 =$

2.
$$\begin{array}{r} 279 \\ 637 \\ 738 \\ 457 \\ 962 \\ + \ 444 \\ \hline \end{array}$$

3.
$$\begin{array}{r} 4{,}789 \\ 6{,}837 \\ 5{,}393 \\ +4{,}579 \\ \hline \end{array}$$

4.
$$\begin{array}{r} 569{,}397 \\ + \ 379{,}688 \\ \hline \end{array}$$

Subtract and check:

5.
$$\begin{array}{r} 4{,}131 \\ - \ 953 \\ \hline \end{array}$$

6.
$$\begin{array}{r} -38{,}332 \\ -19{,}049 \\ \hline \end{array}$$

7.
$$\begin{array}{r} 402{,}302 \\ -91{,}559 \\ \hline \end{array}$$

8.
$$\begin{array}{r} 40{,}050 \\ -8{,}592 \\ \hline \end{array}$$

Multiply and check:

9.
$$\begin{array}{r} 238 \\ \times \ 79 \\ \hline \end{array}$$

10.
$$\begin{array}{r} 5{,}698 \\ \times \ 209 \\ \hline \end{array}$$

11.
$$\begin{array}{r} 8{,}030 \\ \times \ 690 \\ \hline \end{array}$$

12. $409 \times 8{,}800 =$

Divide and check:

13. $4\sqrt{7{,}604}$ **14.** $9\sqrt{27{,}648}$ **15.** $26\sqrt{157{,}664}$ **16.** $54\sqrt{96{,}498}$

8-5. Basic Math Problems: Fractions

Remember to reduce all answers to lowest terms.

Add:

1.
$$\begin{array}{r} \frac{3}{4} \\ + \ \frac{5}{8} \\ \hline \end{array}$$

2.
$$\begin{array}{r} \frac{2}{5} \\ \frac{1}{2} \\ + \ \frac{7}{10} \\ \hline \end{array}$$

3.
$$\begin{array}{r} 2\frac{2}{3} \\ 5\frac{1}{4} \\ + \ 3\frac{5}{6} \\ \hline \end{array}$$

4.
$$\begin{array}{r} 5\frac{2}{3} \\ + \ \frac{4}{5} \\ \hline \end{array}$$

Subtract:

5.
$$\begin{array}{r} 12\frac{5}{8} \\ -4\frac{1}{2} \\ \hline \end{array}$$

6.
$$\begin{array}{r} 10 \\ -3\frac{3}{4} \\ \hline \end{array}$$

7.
$$\begin{array}{r} 6\frac{2}{3} \\ -2\frac{3}{4} \\ \hline \end{array}$$

8.
$$\begin{array}{r} 18\frac{1}{6} \\ -9\frac{2}{3} \\ \hline \end{array}$$

Multiply:

9. $\frac{2}{3} \times \frac{5}{6} =$ **10.** $\frac{3}{4} \times 14 =$ **11.** $\frac{3}{5} \times \frac{10}{12} =$

12. $4 \times \frac{1}{3} =$ **13.** $4\frac{3}{5} \times 8 =$ **14.** $2\frac{3}{4} \times 1\frac{1}{3} =$

Divide:

15. $\frac{5}{6} \div \frac{2}{3} =$ **16.** $\frac{3}{5} \div \frac{10}{12} =$ **17.** $\frac{2}{3} \div 12 =$

18. $9 \div \frac{3}{4} =$ **19.** $5\frac{1}{6} \div \frac{1}{3} =$ **20.** $7\frac{2}{5} \div \frac{1}{4} =$

8-6. Basic Math Problems: Decimals

Add and check:

1. $5.4 + 7.68 + 19 + 2.2 =$

2.
$$\begin{array}{r} 64.09 \\ 0.85 \\ 1.07 \\ 10 \\ 3.25 \\ \hline \end{array}$$

3. $81.51 + 65.52 + 11.5 + 14 =$

4. $\$55.15 + \$0.75 + \$10 + \$12.37 =$

8-6. Basic Math Problems: Decimals (continued)

Subtract and check:

5.   32
   − 6.17

6.   23.5
   − 6.01

7.   72.519
   − 2.36

8.   632 − 2.79 =

Multiply and check:

9.   23.3
   × 4.3

10.   1.849
   × 37

11.   5.766
   ×8.01

12.   47.3 × 0.095 =

Divide and check:

13. $8.3\sqrt{.0415}$

14. $1.08\sqrt{5.8968}$

15. $3.22\sqrt{64,400}$

16. $9.06\sqrt{832.2516}$

8-7. Basic Math Problems: Feet and Inches
Perform the Indicated Operations

1.   2 ft 6 in + 13 ft 9 in
2.   18 ft 5¼ in + 3 ft 2⅛ in
3.   141 ft 2⅛ in + 9 ft 4½ in
4.   35 in 3¾ in + 9⅜ in
5.   131 ft 2 in − 8 ft 6 in
6.   83 ft 4 in − 19 ft 2⅛ in
7.   99 ft ½ in − 35 ft 3 in
8.   1 ft 2 in − 6⅛ in
9.   18 ft 3 in × 5
10.   183 ft 6⅛ in × 8
11.   16 ft 8¼ in × 3
12.   8⅜ in × 4
13.   8 ft 9¼ in ÷ 4
14.   13 ft 3⅛ in ÷ 8

8-8. Degrees, Minutes, and Seconds

Add the following:

1.   25° 6' 18"
   + 4° 10' 25"

2.   9° 35' 5"
   +8° 31' 59"

3.   19° 2' 1"
   0° 3' 59"
   +2121° 22' 1"

Subtract the following:

4.   23° 10' 21"
   − 1° 5' 20"

5.   18° 22' 1"
   − 5° 23' 5"

6.   26° ' 8"
   − 7° 10' 12"

Multiply the following:

7.   6° 5' 8"
   × 4

8.   21° 0' 5"
   × 6

9.   18° 23' 21"
   × 8

Divide the following:

10. $5\sqrt{26° 30' 25"}$

8-9. Squares, Logs, and Square Roots using <u>Smoley's</u>

Use *Smoley's* to look up or perform operations below:

1. Square of 6 ft 3 in
2. Square of 13 ft 3½ in
3. Square of 181 ft 3½ in
4. Square of 8¾ in
5. Square root of 125.26
6. Square root of 18.621
7. Square root of 2,516.00
8. Square root of 8,598.6984
9. Log of 92 ft 7 in
10. Log of 35 ft 4⁷/₁₆
11. Log of 11 ft 9¼ in
12. Log of 5 ft 3⅛ in
13. Feet and inches for 0.59292   log
14. Feet and inches for 0.50060   log
15. Feet and inches for 1.29003   log
16. Feet and inches for 1.46859   log
17. Log sine of 27° 13 ft
18. Log tangent of 35° 6 ft
19. Log cosine of 15° 8 ft
20. Natural log sine of 15° 8 ft
21. Natural log tangent of 41° 18 ft
22. Natural log cosine of 36° 43 ft
23. Find degrees for common log sine 9.84566
24. Find degrees for common log cosine 9.60931
25. Find degree for natural log sine of 1,146
26. Find degrees for natural log sine of 7,660
27. Find degrees for natural log tangent of 1.4835

Use D.E. section of *Smoley's* to find decimal equivalent of numbers below:

28. 8½ in
29. 11⅞ in
30. 5⅜ in
31. 1 ft 6⅜ in
32. 13 ft 3⅛ in
33. 21 ft 6¼ in

8-10. Use Fig. 8-39 for this Exercise

8-11. Use Fig. 8-40 for this Exercise

Nine

Plan Elevations and Fabrication Drawings

This chapter deals primarily with orthographic projection as it relates to piping plans, elevations, and fabrication drawings. Orthographic projection was briefly discussed in Chap. 2 and it would perhaps be advisable at this point to refer back for a brief review. Orthographic projection is used in all types of drafting, including architectural, structural, mechanical, and piping. Drafting standards and conventional practices are universal, making it possible for people all over the world to have a common communication medium. Individual areas of drafting differ somewhat in their terminology and technical information, but the basics of orthographic projection are the same.

After completion of this unit the student should be able to visualize views of pipe and fittings and supply missing views to pipe-layout problems. The student should also be able to visually cut sectional views through plan views using plan and elevations drawings and coordinate dimensions. Notes, dimensions, and other necessary information to complete plan and elevation orthographic drawings should be included.

Orthographic views are used extensively for plans, elevations, and sections as well as on spool drawings when a company prefers orthographic over isometric spools.

EXPLANATION OF ORTHOGRAPHIC PROJECTION

Orthographic projection is the drawing of views of surfaces or planes at right angles to each other. The top, front, right side, left side, bottom, and back are generally called the *principal views*. In basic drawing, these views of an object are very easy to visualize. Piping visualization is a bit more complicated, in that pipe runs go in many directions and often are behind or in front of other pipe runs. Orthographic views are difficult to dimension at times due to this problem of congestion of views.

Orthographic drawings (or *plan and elevation drawings*, as they are also called) are drawings of process pipe systems. They are usually drawn to a scale of ⅜ in = 1 ft 0 in. Orthographics on spool drawings, however, are usually not drawn to scale. *Spools* are individual sections that make up long runs of pipe in a piping system.

The information shown on spools is for fabrication purposes and for field erection. Spool drawings are not drawn to any scale.

INFORMATION SHOWN ON ORTHOGRAPHIC PIPE PLANS

Each pipe company has its own procedure and requires specific information on its pipe drawings. There are several items that are common to almost all pipe drawings:

Equipment Identification: Each piece of equipment, such as a heat exchanger or pressure vessel, must be located and identified by each company's system of marking.

Pipe: Pipe is dimensioned to its centerlines. A complete pipeline designation system is used to identify each pipe's size, specification, line number, location, and contents.

North Arrow: Each sheet has a north arrow symbol to help correlate the sheets and keep pipelines running in the right direction.

Valve Identification: Valves are normally called out with a particular identification number, such as size and code, and each run of pipe is identified with a flow direction arrow.

Piece Marks: Piece marks are used to indicate how a pipe run is broken into spools of shipable lengths that can be fabricated in a shop.

Sectional Views: Sectional views are indicated on each plan sheet. They are numbered so that a sectional view can be traced back to its origin and cross-referenced several sheets later in a set of plans. Section letters and page numbers are used.

Titleblock Information: Titleblock information includes drawing number, dates, title, revision data, company name, drafter, engineer, and reference sheets. This information is often a vital factor in proper coordination of drawings and engineering processes.

Miscellaneous Information: This consists of many other items that may be necessary in a specific situation. The drafter should become familiar with all such information to become more effective. To become familiar, it is helpful just to study plans and ask questions concerning data shown.

COORDINATE AND ELEVATION DIMENSION SYSTEMS

On the plan view of an orthographic set of plans, a north arrow indicates the north direction. In relation to this north arrow there are north-south coordinates and east-west coordinates (see Fig. 9-1). Think of south as 0. As one goes farther and farther away from south toward north, the north coordinates get larger (see Fig. 9-2). The same principle is true on the east-west coordinate. Think of west as 0; as one goes toward the east, the east coordinate gets larger (see Fig. 9-3).

Elevations start at the lowest point on a pipe system. That point is usually arbitrarily assigned an elevation of 100 ft 0 in. Going up from that point the elevations get larger (see Fig. 9-4). Elevation dimensions are up and down dimensions. Coordinate dimensions are north-south or east-west.

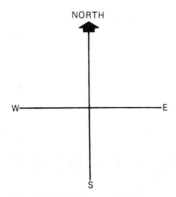

Fig. 9-1 North coordinates.

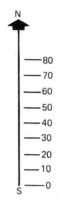

Fig. 9-2 North-South coordinates.

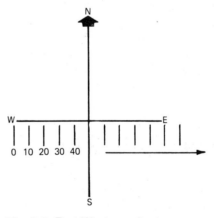

Fig. 9-3 East-West coordinates.

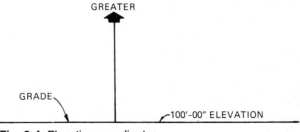

Fig. 9-4 Elevation coordinates.

113

RULES IN DIMENSIONING AN ORTHOGRAPHIC VIEW

The basic principles of dimensioning orthographic views apply in piping. Some general considerations will be mentioned at this point.

A. Most dimensions are given from centerline of pipe to centerline of pipe.

B. Keep dimensions above line and reading from lower right-hand corner. (See Fig. 9-6.)

C. Dimensions should read in feet and inches for 1 ft 0 in and over, for example, 25 ft 6 in or 3 ft 3½ in, or 4 ft 0 in.

D. Place dimensions in a position where they most clearly describe the part being dimensioned.

E. Avoid duplication of dimensions.

F. Do not tie equipment together with dimensions; locate equipment with coordinates.

G. Do not dimension fitting makeup runs.

H. There should be considerable contrast between the pipelines and the lighter dimension and extension lines.

HOW TO DETERMINE SPOOL BEGINNINGS AND ENDINGS

Perhaps the simplest spool run is a piece of pipe with a flange on each end. Spool runs get much more complex but generally are limited by one or more of the following factors:

A. The spool run begins and ends with a flange (see Fig. 9-5).

B. The spool run begins with a flange, but because the distance to the next flange is so great, the spool may end with a plain- or bevel-end piece of pipe (see Fig. 9-6).

C. The method used to transport spools governs the length of each leg of the spool. If the legs are too long, the spool may be broken at the nearest possible fitting so that it can be transported (see Fig. 9-7) by truck, train, or barge.

Spool numbering begins upstream and continues downstream with successive dash numbers for spool identifications. Each spool has a separate callout but is tied back to the main line. Figure 9-8 is a typical isometric layout of a complete pipe run for line 020–90–6″. The spools begin at the lower right, which is upstream. Spool 020–90–6″–1 begins there, and other spools progress around to the left. Spools are separated with field weld notations. It is necessary to draw each spool either as an orthographic or as an isometric spool. Information such as sizes, dimensions, notes, and bill of material is included so that

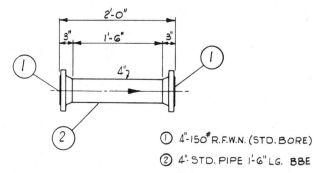

Fig. 9-5 Pipe with two flange spools.

① 4″-150# R.F.W.N. (STD. BORE)
② 4″ STD. PIPE 1′-6″ LG. BBE

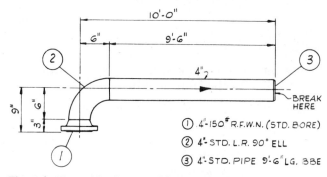

Fig. 9-6 Pipe with elbow and flange spool.

① 4″-150# R.F.W.N. (STD. BORE)
② 4″ STD. L.R. 90° ELL
③ 4″ STD. PIPE 9′-6″ LG. BBE

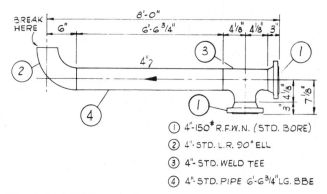

Fig. 9-7 Multifitting spool.

① 4″-150# R.F.W.N. (STD. BORE)
② 4″ STD. L.R. 90° ELL
③ 4″ STD. WELD TEE
④ 4″ STD. PIPE 6′-6¾″ LG. BBE

the weld fabricators can properly assemble each spool from 020–90–6″–1 to 020–90–6″–5. (For more information about fabrication and spooling see Chap. 11.)

THE SYSTEM OF CROSS-REFERENCING DRAWINGS AND VIEWS

To completely detail even an apparently simple process piping system requires dozens of drawings, specification sheets, and bills of material to include all information necessary. It becomes quite obvious that a detail drafter needs a cross-reference system of drawings and views in order to refer from one drawing to another and from a view on one drawing to a view on another drawing.

Each company may have its own system, but basically they all are very similar.

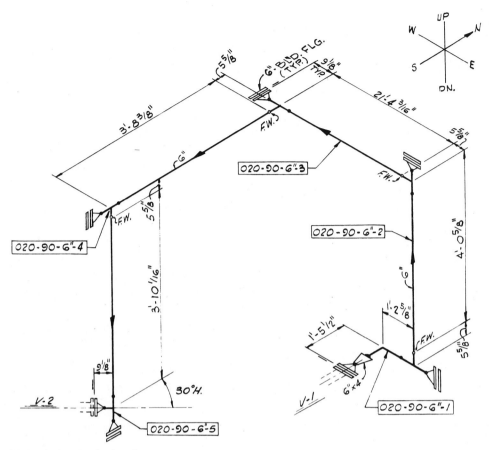

Fig. 9-8 Isometric drawing.

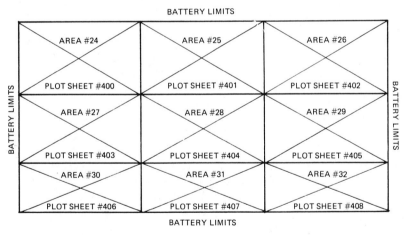

Fig. 9-9 Battery limits.

First, a refinery is divided into areas and each area is numbered (see Fig. 9-9). These numbers become a part of the pipe number as well as the sheet number. This is sometimes called *indexing* a plant facility.

When a total plot is very large, it may involve using plot sheets 400 through 408 to completely lay it out. Each of the plots may be divided into sheets drawn at a scale of ⅜ in = 1 ft 0 in.

Each sheet should have a drawing number referring to its area (see Fig. 9-10, p. 116). A sample sheet number may be 24–120–1. The number 24 stands for the area of the refinery from which the drawing is taken. The number 120 stands for the individual drawing sheet number. The number 1 stands for the latest revision series that has been incorporated on the drawing. See Fig. 9-11 (p. 116) for a portion of an

		AREA 24 LIMITS
SHEET #120	SHEET #121	SHEET #122
SHEET #123	SHEET #123	SHEET #124
SHEET #125	SHEET #126	SHEET #127

Fig. 9-10 Area limits.

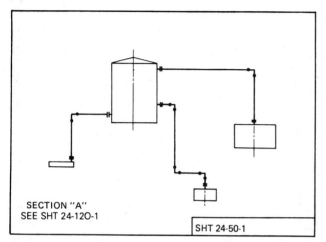

Fig. 9-11 Drawing grid system taken from a plot plan.

actual drawing grid system. If a section view were to be cut through a part of the pipe plan on sheet 24–120–1, it could be drawn on that same sheet if room were available. But if the section were too large, it would be drawn on a separate sheet with its own drawing number, such as 24–50–1. The sectional view on sheet 24–50–1 is referred back to sheet 24–120–1 as its source. The section view could be identified as section A (see Fig. 9-11). When a line originates in area 24 it will bear that number as a part of its total line identification, such as line 8″–24–351–SL. The number 8 refers to the size of line, 24 to the area the line originates from, 351 to the individual line number. SL is the line specification.

After a period of time the drafter should become familiar with the relationship of plot plans, grid sheets, plan drawings, and sectional views. A thorough understanding of the pipeline designation system for each pipeline is necessary for proper understanding of the total set of working drawings. This understanding of the pipe reference system will better qualify the drafter to refer from sheet to sheet and in the process to pull information from sections, details, and line callouts necessary to complete a job.

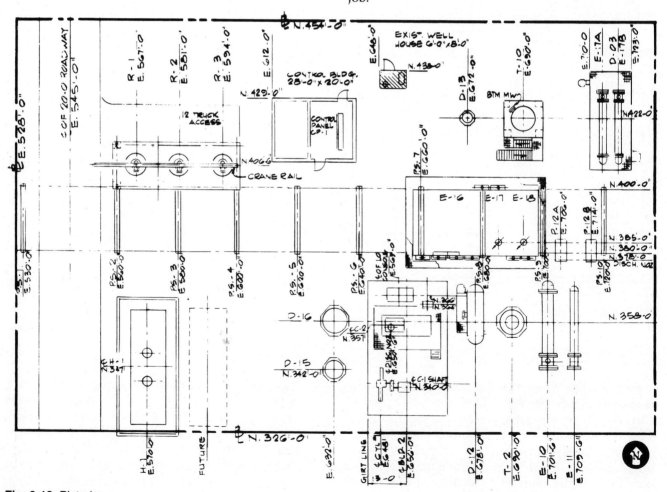

Fig. 9-12 Plot plan.

CONCLUSION

To become skilled as a pipe drafter requires a thorough understanding of all the phases of plan development from the flow diagram stage through plot plans, elevation, plan views, and isometrics of individual lines, to erection drawings and finally to the spool drawing.

These working drawings and the specifications comprise the total picture of a complete job. A seasoned drafter with broad and diversified experience should be able to function at any level of difficulty from the spools to pipe-design flow diagrams. This chapter is a brief introduction to pipe blueprint interpretation.

EXERCISES

9-1. Define the following terms:

1. Orthographic projection
2. Elevation
3. Section
4. Plan view
5. Auxiliary view
6. Projection line
7. Rotating the view
8. Spool drawing
9. Scale
10. Flow direction
11. Piece mark
12. Revision
13. Field weld
14. Coordinate location
15. North arrow
16. Battery limits
17. Match lines
18. Area
19. Sheet number
20. Line designation

9-2. Match the terms and abbreviations:

1. elev.   **a.** Auxiliary view
2. aux. view   **b.** Revision
3. proj.   **c.** Piece marks
4. rev.   **d.** Orthographic projection
5. F.W.   **e.** Elevation
6. P.M.   **f.** Projection
7. sht. no.   **g.** Field weld
8. sect.   **h.** Sheet number
9. iso.   **i.** Section
10. ortho. proj. **j.** Plans and elevation
11. P & E   **k.** Centerline
12. C_L **l.** Isometric
13. sc.   **m.** Scale

JOB ASSIGNMENTS

9-3. Redraw Fig. 9-7 (spool drawing) to scale and include bill of material.

9-4. Redraw Fig. 9-8 (isometric line drawing) at no scale. Include dimensions and notes, and compile a bill of material.

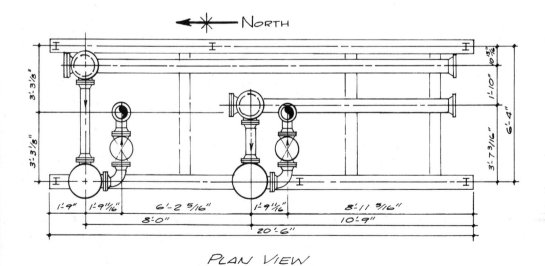

Fig. 9-13 Plan view of an actual job.

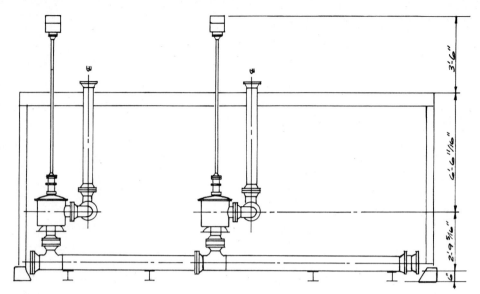

Fig. 9-14 Actual elevation taken from plan view in **Fig. 9-13.**

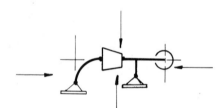

Fig. 9-15

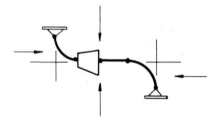

Fig. 9-16

9-5. Redraw Fig. 9-13 (plan view) to the scale of ⅜ in = 1 ft 0 in. Draw it as a single-line pipe drawing.

9-6. Redraw Fig. 9-13 (plan view) as a two-line pipe drawing to the scale of ⅜ in = 1 ft 0 in.

9-7. Redraw Fig. 9-14 (section) as a two-line pipe drawing.

9-8. Redraw Fig. 9-14 (section) as a single-line pipe drawing.

9-9. Orthographic Pipe Exercises

A. Use dividers or scale and redraw Fig. 9-15 on a sheet of 8½ × 11-in paper. Draw a top, right side, left side, and bottom.

B. Redraw Fig. 9-16 as described in question A.

C. Redraw Fig. 9-17 as described in question A.

D. Redraw Fig. 9-18 as described in question A.

E. Redraw Fig. 9-19 as a double-line plan view.

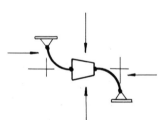

Fig. 9-17

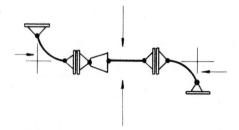

Fig. 9-18

118

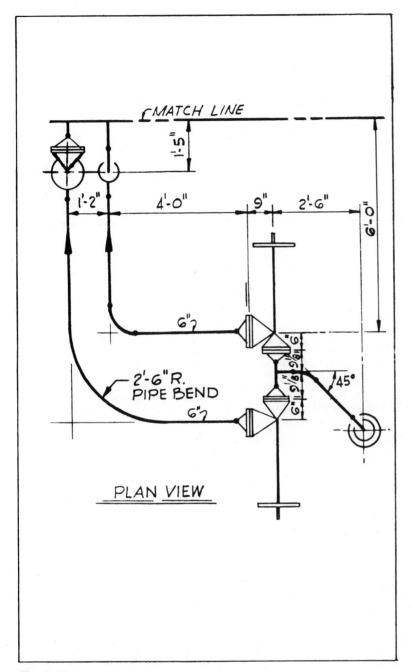

MATCH LINE

1'-5"

1'-2"   4'-0"   9"   2'-6"   6'-0"

6"

2'-6" R.
PIPE BEND

6"

45°

PLAN VIEW

Fig. 9-19

Ten

Isometric Drawings

After becoming familiar with orthographic projection methods, the student will find that in certain cases an understanding of isometric illustration is useful. Isometric drawings are used to illustrate, in picture form, complicated piping arrangements, smaller individual pipe runs, and on an even smaller scale, individual spools or shop fab drawings.

Isometric drawings can be drawn to scale but commonly are drawn at no scale so that crowding and overlapping can be avoided. Isometric drawing lends itself to dimensioning and calling out of individual components. Isometric illustration is a universal technique used quite extensively in several types of drafting, especially pipe drafting.

After completion of this unit the student is expected to be able to visualize pipe, fittings, and valves and to correctly illustrate them in isometric form. The student should be able to pull information from larger orthographic and isometric line drawings; properly orient pipe runs, dimension distances, and sizes of fittings; and select the best method for illustrating the roll of pipe runs.

PRINCIPLES AND APPLICATIONS OF ISOMETRIC VIEWS

All normal piping isometric jobs are within the scope of the isometric system. Its application to jobs that are somewhat complex depends upon the skill of the drafter. Scale is not used, and only that degree of proportion needed to keep the views in balance is maintained. A 30°–60° triangle is used to draw isometric drawings. The main axis is 30°.

The isometric system of drawing is one-plane straight-line projection (called *isometric projection*). The projection shows the object's height, width, and depth in one view. The plane on which the view is projected is called the *front plane* (see Fig. 10-1).

ISOMETRIC AXIS LINES

The projections of the three edges of the cube, which meet at its front corner and make three equal angles of 120°, are called the *isometric axes* (see Fig. 10-1).

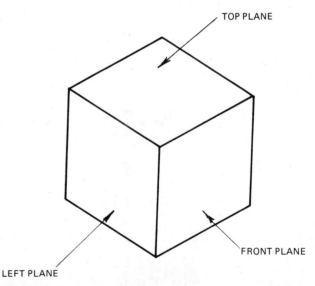

Fig. 10-1 Isometric projection planes.

The planes of the faces of the cube and all planes parallel to them are called *isometric planes*. The principal lines in piping systems are considered as falling in separate isometric planes, which are related plane-to-plane-to-plane dimensioning (see Fig. 10-2).

ISOMETRIC LINES

Any line parallel to an edge of the isometric cube projects parallel to an isometric axis. All vertical lines project parallel to the vertical axis. Only those horizontal lines which are parallel to the edges of the cube project parallel to the horizontal axis. The projections of these lines are called *isometric lines*. Figure 10-3a illustrates the isometric projection of a typical piping installation in which all the lines are parallel to the edges of the cube. The projections are thus isometric lines.

The isometric plane in which a diagonal line lies is indicated by framing the diagonal in either an isometric square (see Fig. 10-4) or an isometric rectangle (see Fig. 10-5). The square or rectangle is drawn parallel to the face of the cube to which the plane is parallel (see Fig. 10-3b, illustrating isometric planes). In scale drawings, the sides of the frames are shown exact scale. In visual isometric drawings (used in detailing pipe isometrics) the frames are proportioned to indicate the approximate size of the angle, except in those cases where the projections must be opened to reveal hidden lines. (Fig. 10-4 shows a true isometric projection of 45° diagonal intersecting the vertical lines at right angles. In this view the diagonal is hidden by the projection of the vertical line. You should avoid this method whenever possible). Diagonals of 45° are framed in isometric rectangles to show pipe without breaking the vertical lines (see Fig. 10-5).

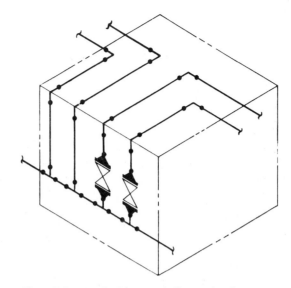

Fig. 10-3a Applied isometric line projection

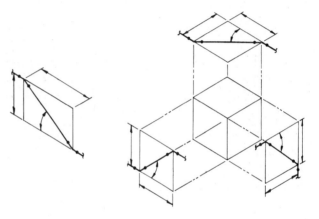

Fig. 10-3b Angular isometric projections.

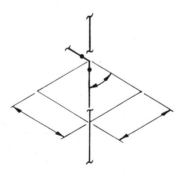

Fig. 10-4 An isometric square.

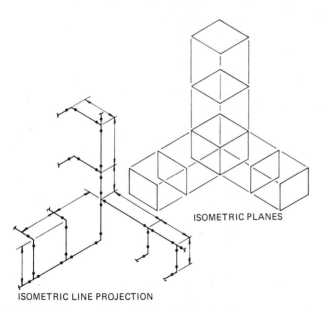

ISOMETRIC PLANES

ISOMETRIC LINE PROJECTION

Fig. 10-2 Isometric projection of lines in planes.

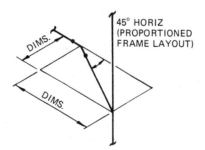

45° HORIZ
(PROPORTIONED FRAME LAYOUT)

DIMS.

DIMS.

Fig. 10-5 A proportioned isometric frame.

121

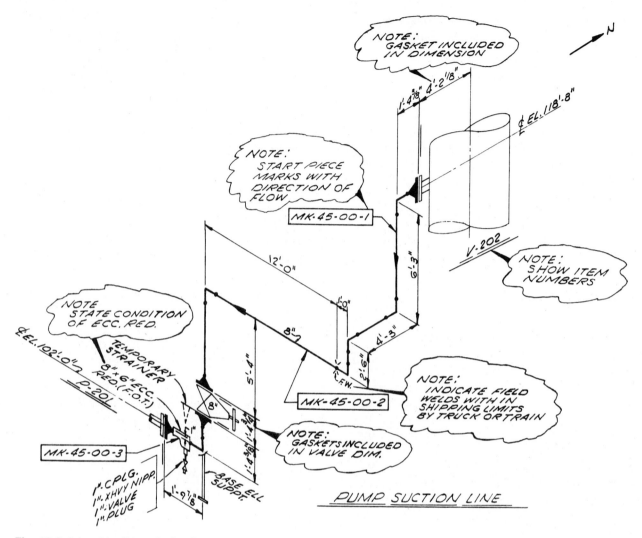

Fig. 10-6 Isometric dimensioning factors.

DIMENSIONING AN ISOMETRIC DRAWING

The dimensions that go on an isometric drawing are determined from specifications, manufacturer standards, and the physical conditions being designed for.

Each isometric is drawn from origin to terminus and is complete with flow direction and line size in two or three places. Coordinates and elevation dimensions on nozzles, rack runs, valve handwheel orientations, and any off-spec items such as 300-lb flanges in a 150-lb specification line should be included. Fabricating dimensions, equipment item numbers to which pipe connects, and "hash marks" for each gasket should also be included.

On the isometric drawings in Figs. 10-6, 10-7, and 10-8a and 10-8b (p. 124) a few of the dimensioning factors are illustrated. Study the drawings and pay close attention to the items noted.

REPRESENTATION OF FITTINGS, VALVES, AND PIPE IN ISOMETRIC

Fittings and valves are represented schematically on isometric drawings just as they are on orthographic drawings. The problem results when correct direction and angles must be accounted for. By practice and observation the student can become very skilled at illustrating a pipe run in isometric.

Some basic components are illustrated on the following pages. Study each and then practice drawing it in each of the planes of the isometric axis. For example, the 90° elbow can be shown in several positions (see Fig. 10-9, p. 124). A short- or long-radius 90° elbow could appear at any of the corners of the isometric plane. Or there could be a welded, flanged, threaded, socket-weld, or other type of connection. Although most fittings are welded, it is very easy to substitute the appropriate symbols to represent other types of fittings.

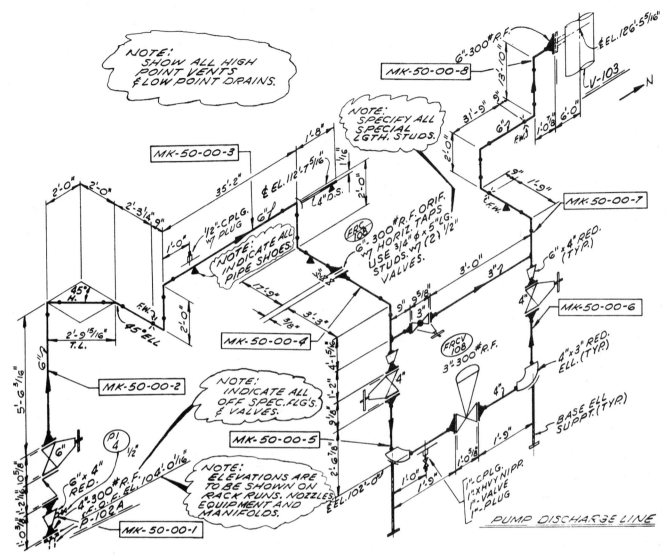

Fig. 10-7 Isometric dimensioning factors.

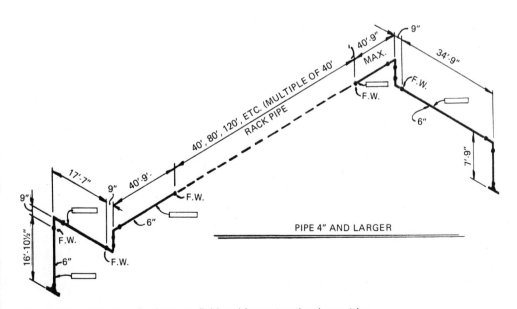

Fig. 10-8a How to split-shop vs. field-weld construction isometrics.

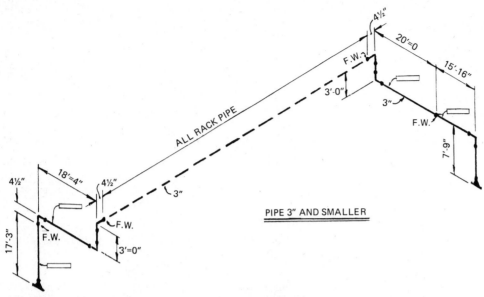

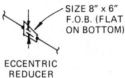

PIPE 3" AND SMALLER

Fig. 10-8*b* How to split-shop vs. field-weld construction isometrics.

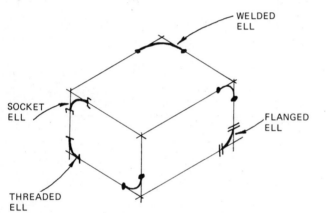

Fig. 10-9 Samples of isometric positions for elbows.

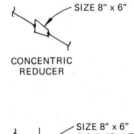

CONCENTRIC
REDUCER

ECCENTRIC
REDUCER

Fig. 10-11 Straight line reducers in isometric drawings.

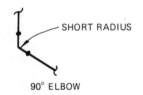

SHORT RADIUS

90° ELBOW

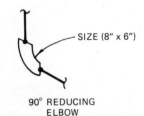

SIZE (8" x 6")

90° REDUCING
ELBOW

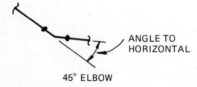

ANGLE TO
HORIZONTAL

45° ELBOW

Fig. 10-10 Representation of elbows in isometric drawings.

REDUCING
TEE

STRAIGHT
TEE

Fig. 10-12 Isometric tees.

To save space, all the positions of each of the isometric fittings are not shown, but the student should practice using a box, as shown in Fig. 10-9, to place various types of fittings in their proper locations.

The ability to draw fittings, pipe, and valves in isometric can be learned by practice if Figs. 10-10 to 10-20 are used as guides for representation. Each depicts a type of fitting. It is impossible to show all the positions and types of welds for each fitting; however, nearly all welded fittings and the more common positions for each fitting are shown.

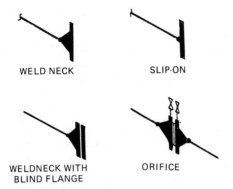

WELD NECK SLIP-ON

WELDNECK WITH BLIND FLANGE ORIFICE

Fig. 10-13 Isometric flanges.

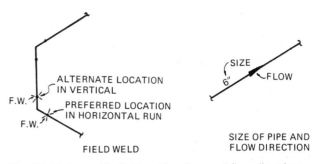

ALTERNATE LOCATION IN VERTICAL

F.W.

PREFERRED LOCATION IN HORIZONTAL RUN

F.W.

FIELD WELD

SIZE 6" FLOW

SIZE OF PIPE AND FLOW DIRECTION

Fig. 10-16 Isometric field welds, size, and flow directions.

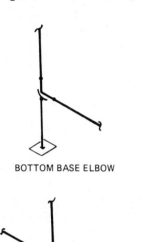

BOTTOM BASE ELBOW

DUMMY LEG

SIDE BASE ELBOW

Fig. 10-14 Isometric elbow and dummy supports.

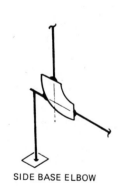

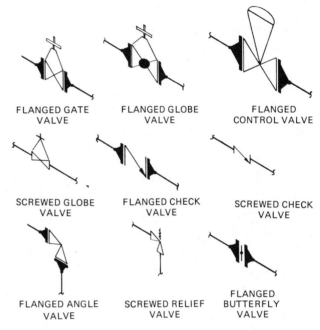

FLANGED GATE VALVE FLANGED GLOBE VALVE FLANGED CONTROL VALVE

SCREWED GLOBE VALVE FLANGED CHECK VALVE SCREWED CHECK VALVE

FLANGED ANGLE VALVE SCREWED RELIEF VALVE FLANGED BUTTERFLY VALVE

Fig. 10-17 Isometric valve symbols.

COUPLING WELD WELDOLET THREADOLET SOCOLET STUB-IN

Fig. 10-18 Isometric branch line connections.

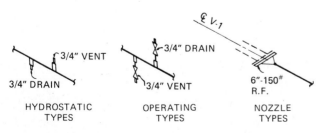

3/4" VENT

3/4" DRAIN

HYDROSTATIC TYPES

3/4" DRAIN

3/4" VENT

OPERATING TYPES

℄ V.1

6"-150# R.F.

NOZZLE TYPES

Fig. 10-15 Isometric vents, drains, and nozzles.

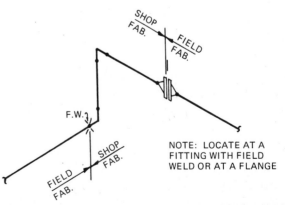

SHOP FAB.

FIELD FAB.

F.W.

SHOP FAB.

FIELD FAB.

NOTE: LOCATE AT A FITTING WITH FIELD WELD OR AT A FLANGE

Fig. 10-19 Isometric representation of shop fabrication and field fabrication.

NORTH ARROW LOCATIONS

The location of the north arrow on isometric as well as plan and elevation drawings is a critical as well as informative part of the drawings. Each line as it is taken from a plan view must be positioned on the isometric so that the line is running in correct relation to its direction on the plan view. (see Fig. 10-21). Basically, the pipe taken from the plant could run in any direction—north, south, east, west, up, down, or at any angle (see Figs. 10-22a and b). The pipe drafter should develop the ability to turn a dimensional plan view into a nondimensional isometric with all fittings, valves, and pipe shown in their correct manner and running in the right north arrow orientation.

On all piping isometrics, the north arrow should point up and to the right-hand corner of the drawing to help the student in drawing an isometric line. The student should be able to visualize the line in three dimensions. One aid to the student is a strand of electrical wire. The wire can be bent to the configuration of the line. After the student has the configuration, the wire can be straightened and ready to form in a new isometric. Another useful and readily available tool for model building is smoking pipe cleaners. They are available in most drugstores.

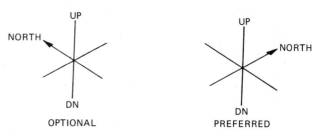

Fig. 10-22a North arrow orientation.

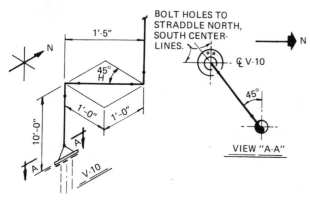

Fig. 10-20 Isometric bolt hole location methods.

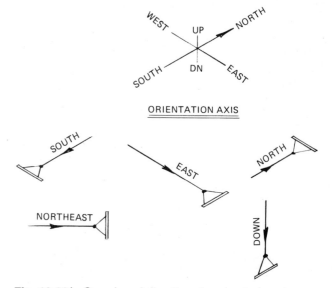

Fig. 10-22b Samples of directions for pipe to travel.

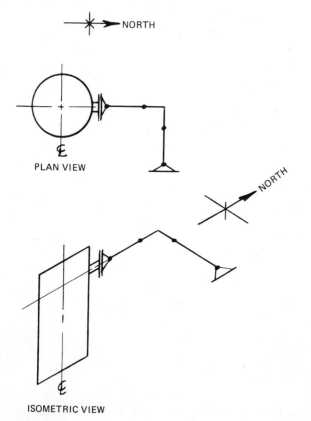

Fig. 10-21 North arrow plan and elevation comparison.

EXERCISES

10-1. Define the following terms:

1. Isometric
2. Isometric axis line
3. Optical illusion
4. Isometric planes
5. Frame
6. Crate
7. Nonisometric
8. Concentric

9. Eccentric
10. Base ell
11. Dummy support
12. Nozzle
13. Vent
14. Drain
15. Field weld
16. Flow arrow
17. Coupling
18. Stub in
19. Field fab
20. Shop fab
21. North arrow
22. Bolt holes straddle

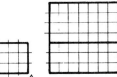
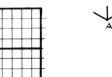

Fig. 10-23

10-2. Match the terms and abbreviations:

1. iso.	**a.** Horizontal
2. proj.	**b.** Isometric
3. pl.	**c.** Flat on top
4. horiz.	**d.** Projection
5. vert.	**e.** Coupling
6. F.O.T.	**f.** Plane
7. cplg.	**g.** Reducer
8. red.	**h.** Vertical
9. mk.	**i.** Mark
10. el.	**j.** Elevation
11. C$_L$	**k.** Field weld
12. F.W.	**l.** True length
13. T.L.	**m.** With
14. w/	**n.** Nozzle
15. conc.	**o.** Centerline
16. F.O.B.	**p.** Concentric
17. noz.	**q.** Flat on bottom

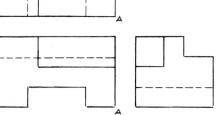

Fig. 10-24

Fig. 10-25

JOB ASSIGNMENT

10-3. Draw the three views in Fig. 10-23 as an isometric.

10-4. Draw the three views in Fig. 10-24 as an isometric.

10-5. Draw the three views in Fig. 10-25 as an isometric.

10-6. Draw the three views in Fig. 10-26 as an isometric.

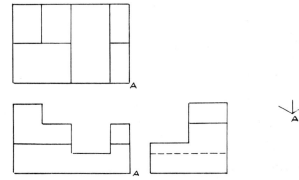

Fig. 10-26

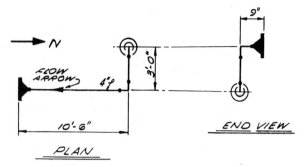

Fig. 10-27

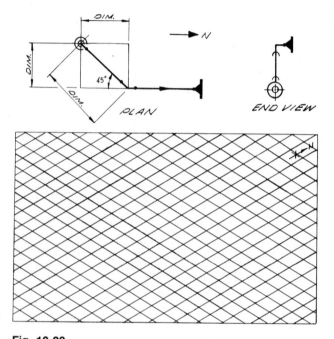

Fig. 10-28

10-7. Convert orthographic pipe views to isometric views: The problem is drawn in orthographic plan and end view. Draw an isometric with the north arrow to the upper right-hand corner of the paper. Use the north arrow shown in the plan to orientate the pipe in isometric. The line is 4″ pipe with 150-lb weld-neck flanges and 4″ long-radius ells (see Fig. 10-27).

10-8. Convert the orthographic views in Fig. 10-28 to an isometric view.

10-9. Convert the orthographic views in Fig. 10-29 to an isometric view.

10-10. Convert the orthographic views in Fig. 10-30 to an isometric view.

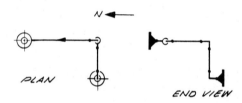

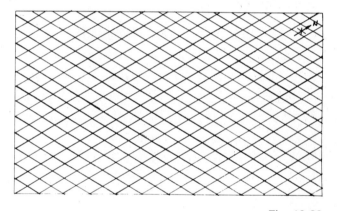

Fig. 10-29

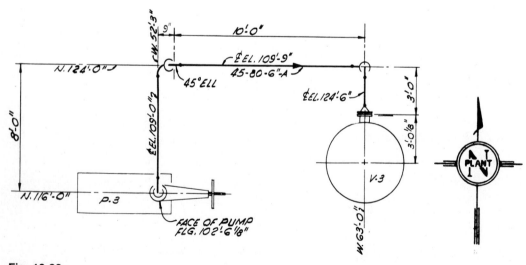

Fig. 10-30

10-11. Convert the orthographic views in Fig. 10-31 to an isometric view.

10-12. Draw section A-A in Fig. 10-32 as an isometric. (Isometrics are not drawn to scale.)

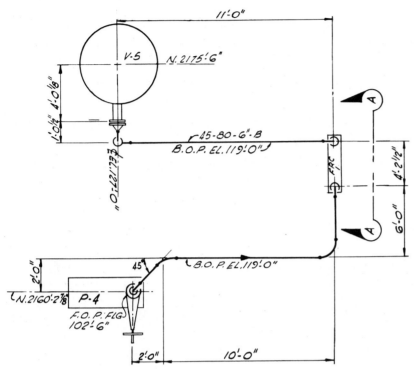

Fig. 10-31

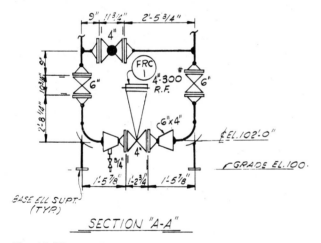

SECTION "A-A"

Fig. 10-32

Eleven

Pipe Fabrication and Spooling

The increasing use of higher pressures and temperatures has elevated the art of designing, erecting, and purchasing fabricated piping. Engineering companies realize that the purchase of piping requires the same careful thought and attention to details as does the selection of steam generators, refinery cracking stills, valves, and other major items of modern plants.

The production and development of fabricated piping have gone hand in hand with the constant improvement in valves, fittings, and allied articles. For this reason, materials of uniform design, workmanship, and performance have been developed.

The nature of fabricated piping precludes any attempt to list completely the variety of materials and operations available in a modern fabricating shop. The special and peculiar requirements of industry and its process present new problems daily which must be solved.

After completion of this unit the student should be familiar with pipe-fabrication methods both in the shop and in the field. The implementation of spool drawings (single-line and double-line) will be emphasized as the basic process of pipe erection.

Orthographic as well as isometric spools have a particular application, and each piece of pipe, valves, and fitting must be compiled for each spool on a bill of material.

Flow sheets, plot plans, elevations, isometrics, and all the preceding paperwork lead to the actual fabrication of spool sections, the end result that makes plant erection possible.

PIPE FABRICATION

After piping drawings have been prepared, piping materials must be procured and fabricated for erection. To *fabricate* piping means to assemble pieces, such as elbows, tees, and flanges, into sections which can be fitted together into the finished plant complete with pipe, flanges, fittings, valves, and all the other equipment.

TWO TYPES OF PIPE FABRICATION

The term *fabrication* is used in piping most commonly when referring to larger size pipe (3" and above) which is joined by welded or flanged joints. Because of the size and importance of such lines and the permanent nature of a welded joint, sections of this pipe must be assembled and welded with a degree of precision that requires careful layout of work, control of welding operations, the use of jigs and templates, and more precise tools than those necessary for threaded pipe. These conditions suggest work at ground level and in an organized shop-type atmosphere found in a permanent pipe-fabricating shop or, at times, in a well-planned temporary field shop.

FABRICATION OF THREADED PIPE

All threaded-joint (screwed) piping and small compression-joint tubing are fitted in the field. In these small sizes the piping runs are measured, cut, threaded, and fitted to the equipment. Although this piping is dimensioned on drawings, it is practical for the pipe to be fitted in the field since pipe 2 to 2½″ or smaller is easily handled by a cut-and-fit technique. Variations or tolerances within practical limits can be established by notes on the drawings, rather than by precise dimensions between each fitting. This is the usual procedure followed in plumbing practice where exact dimensions are rarely shown on drawings. Because of the minor space needs of small piping, precise location is not important.

SHOP FABRICATION FACTORS

Portable tools and equipment are now manufactured that permit most shop fabrication methods to be duplicated in the field, except heat-treating of large tonnages of fabricated pipe.

The factors which determine the extent of shop or field fabrication are varied and may be governed by the erector's previous experiences, piping size and type, and the erector's aggreements with organized labor in the area. If all piping is field-fabricated, the erector must maintain personnel experienced in the usual production-line procedures followed in fabricating shops. In addition, tools and equipment ordinarily found only in shops will be required. A more extensive field organization is also needed for material control, material handling, and the performance of certain technical operations. Additional shelter in the form of temporary sheds or buildings is necessary for part of the fabricating operations, for warehousing of certain materials, and for housing the added technical and clerical personnel. Erectors who perform many large jobs each year and maintain a wide assortment of equipment and materials often find favorable conditions for complete field fabrication. There is, however, no general rule which is rigidly followed. This is a competitive field, and shop fabricators are constantly proving their value by providing competent workmanship at low cost. Thus, the varied experiences of erectors with shop and field fabrication will often have a strong influence on their decision.

Piping size and type can have a significant influence on whether piping is shop- or field-fabricated. Alloy piping often requires involved techniques for heat-treating which are more difficult to perform in the field than in the shop. There are times, however, when the alloy piping is so extensive that it may be economical to provide the necessary apparatus in the field, since a certain amount of field fabrication must be performed, and this way the alloy piping will receive the same type of heat treatment as the rest of the pipe. Pipe in 3″ and 4″ sizes is relatively easy to handle and often proves attractive for field fabricating.

Erectors must make agreements with local organized labor, and at times these agreements fix the amount of piping that must be field-fabricated by establishing a break point on the size rather arbitrarily. These agreements are sought by the local unions in order to ensure work for their members, because shop-fabricated pipe may be prepared in some other area of the country.

A piping system, when placed in service, should develop the predetermined results and fulfill all the expectations of a proper design. The installation should be accomplished expeditiously and at a minimum of final erected cost consistent with the use of good materials and workmanship. Obviously, this purpose will be defeated unless each piece, large and small, of both simple and complex piping systems, receives the attention necessary to assure accuracy and a high quality of workmanship. Shop fabrication accomplished this desirable end.

Mechanics in a fabricating shop are employed solely on the production of fabricated piping. Such continuity of practice means that they do not lose their proficiency, particularly important for welding and other complex operations.

Safety is a prime requisite of every piping installation. Shop fabrication assures purchasers that every necessary precaution has been employed to eliminate hidden hazards of piping. A weld may appear excellent on the exterior, yet under the surface it may contain gas pockets or have faulty penetration. A faulty threaded flange joint gives no indication, in its external appearance, of improper threading. These are two of the many hidden hazards of piping.

Shop fabrication requires the use of special machinery, clamps, jigs, and other convenient appliances which are impractical for field service. This equipment permits the alignment and squaring up of all flanged faces and ends of piping, and preserves the accuracy of dimensions and angles—all within the closest practical limits. Such characteristics assure the final fitting of every piece into position without alteration or improper strain, resulting in increased safety and a saving of time and labor. Stress-relieving or heat-treating, when necessary, are readily performed in the shop. All of these advantages are almost impossible to obtain in the field.

Working conditions, almost invariably, are better in a shop than in the field. The shop has facilities for handling heavy or bulky pieces; therefore, each mechanic is working at all times in a comfortable position and need not be concerned about safety.

Shop fabrication assures dependability because the product is tested and proved sound before shipment. Piping fabricated in the field usually must be installed in its final position before it can be tested, and in the event of failure, the removal for replacement or repair proves expensive and causes delay.

Shops are fully equipped with machinery and facilities for the various operations required in the production of a complete line of fabricated piping. Much shop equipment is especially designed to fabricate piping of uniform accuracy and quality of workmanship.

Many forms of jigs and appliances are employed constantly to supplement the effectiveness of the machines and other facilities.

In fabricating shops, the art of bending pipe has been developed to a high degree. Pipe bends of many sizes and shapes are in constant production. Pipe bends have a wide range of usefulness. No two general piping installations are alike: hence, pipe bends are especially suitable for solving the many problems of space requirements. Getting the proper proportions of a pipe bend usually is a matter of bending pipe to the dimensions necessary to fit into the available space. To meet these needs of industry, pipe bending facilities are available; shops are equipped to fabricate pipe bends from a large range of sizes, weights, and kinds of pipe.

PIPE SPOOL DRAWING DEFINED

A *pipe spool drawing* is a unit of piping that is limited in length or size by either shipping dimensions or volume. A spool consists of a one- or two-line drawing with major dimensions as well as subdivided dimensions of individual components. Each component is labeled and later described in a bill of material. The spool becomes a separate drawing complete with all information necessary for the welder to completely assemble the unit (see Figs. 11-1 and 11-2 and the set of plans in the Appendix).

Each spool has mark numbers, and these mark numbers or spool numbers also appear on the piping drawing so that the erection crews can determine the sequence of pieces relative to equipment. Note also that a bill of materials appears on each spool drawing. This enables the fabricator to place the orders for the correct amount of materials and also gives the welder a handy summary of the materials that must be assembled for the particular spool. Thus the spool drawing is a complete description of a unit of piping, but it is a particular kind of description oriented to the people who must purchase and assemble the unit. The mark number is also painted on each finished spool for the convenience of the erection crew. The marks are based on the line numbers appearing on the original piping drawing and indicated under the terms identification.

If piping is large and alloy steel, it is fabricated as completely as possible in order to avoid excessive field welding the field heat-treating. Thus there are many factors that must be evaluated, and the piping designer should always consult with traffic people and especially the project engineer before spooling the major lines.

METHODS OF REPRESENTING A SPOOL DRAWING

Keep in mind that a spool is a portion of an entire run of pipe that must be detailed completely so that the weld fab shop can put it together.

A spool can be drawn in single-line orthographic (see Fig. 11-1) or in isometric (see Fig. 11-2). The method often depends on the company for which the work is being done. Spools can also be drawn double-line orthographic (see Fig. 11-3). Spool drawings are not to scale but are drawn to describe correctly how the pipe and fittings are put together. Valves are not included as a part of a spool but are shipped individually to avoid damage.

IDENTIFICATION OF AN INDIVIDUAL PIPE LINE THROUGH LINE DESIGNATION

Shop-fabricated lines in most cases are from 3" pipe and larger. Each isometric pipeline has field welds located within shipping limits either by truck or train. Each piece of spool has a piece mark, as shown below:

MK — 45 — 101 — 1
Unit no.   line no.   numerical sequence

When pipe is shop-fabricated, each piece is marked with this spool number, which is matched in the field with the numbers appearing on the piping isometric that has been issued for erection. Pipe fabricators furnish prints of their shop drawings called *spool sheets.*

CALLOUTS FOR PIPE, FITTINGS, AND FLANGE BORES FOR BILL OF MATERIAL, NOTES, OR SPECIFICATION SHEETS

The bill of material is just as important as the drawing, and all items must be called out. Where possible use the term *standard weight* (abbreviated std.) or *extra heavy* (X-hvy.) in lieu of *schedule* (sch.) or *wall thickness* (W.). For example:

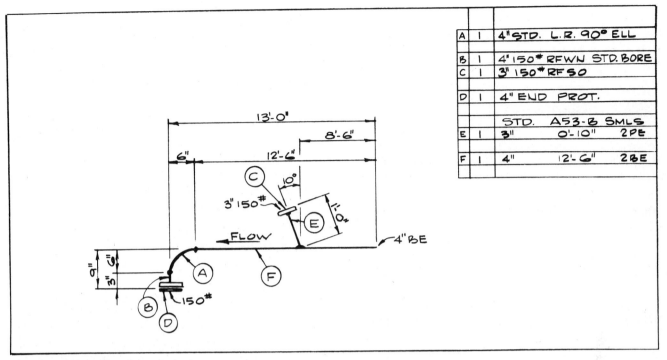

A	1	4" STD. L.R. 90° ELL	
B	1	4' 150# RFWN STD. BORE	
C	1	3" 150# RFSO	
D	1	4" END PROT.	
		STD. A53-B SMLS	
E	1	3"	0'-10"	2PE
F	1	4"	12'-6"	2BE

Fig. 11-1 Single-line pipe spool.

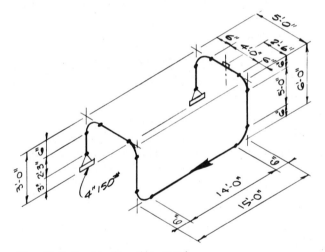

Fig. 11-2 Double-line pipe spool.

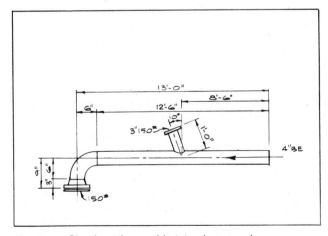

Fig. 11-3 Simple orthographic two-view spool.

12″ std. L.R. 90° ell

10″ X-hvy. 45° ell

10″, 150# W.N. R.F. (std. bore)

In cases where std. and X-hvy. do not apply, use sch. For example:

12″ sch. 40 L.R. 90° ell

10″ sch. 80 45° ell

16″, 300# W.N. R.F. (60 B.)

Note: Schedule is omitted from flange bores.

In special cases where wall thickness does not match standard, extra heavy, or schedule, use the term *wall thickness* (W.). For example:

10″ (0.750″ W.) L.R. 90° ell

14″ (1.000″ W.) 45° ell

16″, 900# W.N. R.F. (14.000″ B.)

The above three notes apply only to 24″ and smaller. All sizes above 24″ are called out by wall thickness.

On flanges larger than 24″, refer to catalog and page number. For example:

30″, 150# W.N. R.F. (29.00″ B.) (T.F. cat. 571, p. 71)

When schedule 10S materials are used, the following rules apply:

A. Use *schedule* for all items 12″ and smaller.

B. On all items 14″ and larger, use *wall thickness*.

133

Determining callout for reducing flanges:

24″ × 34″ O.D. 150# F.F. red. S.O. flange

(24 1⅜″ diameter bolt holes on a 31¾″ bolt circle)

COMPARISON OF ORTHOGRAPHIC AND ISOMETRIC DRAWINGS

In orthographic drawing, straight-line projection is used to show the flanged ell in height, width, and depth. The projections from this position show the front and side view of the flanged ell. The exact outlines of the flanged ell as viewed from front to side are reproduced in two views. They show the flanged ell in shape and in size. On simple shop-fabrication piping, straight orthographic projection is used in two views (see Fig. 11-4).

On complex shop-fabrication piping, where several views are required to show pipe and fittings, the isometric system is used. The isometric system has two advantages which result from the pictorial character of the drawings: The drawings require less time to make and are easier to read than conventional orthographic drawings. Being pictorial in character the drawings show the object much as it would appear to the eye and leave little to be imagined. The views are easy to visualize both before and after drawing. With these advantages the system has several disadvantages that restrict its use for detail purpose. Isometric projection shows the object in three dimensions but obscures many of its details.

Projections of remote lines fall behind projections of near lines, and the views are difficult to dimension. Scale is not used, and only that degree of proportion needed to keep the views in balance is maintained.

All normal pipe detailing jobs are within its scope. Its application to jobs that are near complex depends upon the skill of the drafter. On jobs within its scope it has advantages in drafting and readability over the orthographic system.

BILL OF MATERIAL

The bill of material is just as important as the drawing on a spool sheet, and much care must be taken to properly list and size all the pipe fittings.

Some companies prefer a definite sequence for items to be called out. This makes it easier to find items and prevents them from being overlooked during the material takeoff process (see Fig. 11-5).

Adequate space is usually a problem, so it is imperative to use abbreviations and symbols when possible. The drafter must learn to use and identify many symbols and abbreviations on the drawing, in notes, in the bill of material, and also on the specification sheets.

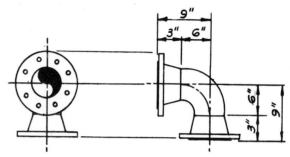

Fig. 11-4 Simple isometric one-line spool.

SPOOLING INFORMATION AND STEPS TO COMPLETING A SPOOL

Spool drawings vary in the shape and size of pipe, but most all include the following information or conform to the following methods of representation:

A. One- or two-line drawings

B. Drawing in isometric or orthographic views

C. Overall dimensions

D. Unit dimensions

E. Bill of material

F. North orientation

G. Line and drawing identification

H. Notes

I. Company information

The drawing of each spool should follow some basic steps for completion to eliminate the possibility of omitting information. Listed below are six very important items. You may want to add to this list or rearrange the order after drawing a few spools.

A. Orient drawing to north

B. Make sketches of line to be spooled

C. Put in fittings and pipe correctly

D. Calculate dimensions and put them in proper location

E. Make out bill of material

F. Fill in titleblock

SUGGESTED STANDARDS FOR NOTES, BILLS OF MATERIAL, AND PIPE-SPOOL DRAWINGS

One of the best ways to get familiar with pipe spools is to study spools that someone else has completed and absorb the methods of representation. Figures 11-6 to 11-12 (pages 135 to 137) are samples of iosmetric spools (drawn single-line). Study each of the spools and pay close attention to the notes and guidelines on each sheet.

QUAN	DESCRIPTION
	L.R. 90° ELLS
	90° RED. ELLS
	S.R. 90° ELLS
	45° ELL
	TEE
	RED. TEE
	CAP
	CONC. RED.
	ECC. RED.
	CONC. SWAGE
	ECC. SWAGE
	L/J STUBS
	150# R.F. W.N.
	150# F.F. W.N.
	150# R.T.J. W.M.
	300# R.F. W.N.
	AND ETC.
	150# R.F. S.O.
	150# F.F. S.O.
	300# AND ETC.
	SCR'D FLANGES
	LAP JOINT FLANGES
	W.N. ORIFICES
	S.O. ORIFICES
	CPLGS.
	WELDOLETS
	SOCKOLETS
	THREDOLETS
	ELBOLETS
	PLATE
	END PROT.
	PIPE

Material Must Be Listed in This Order

List Lowest SCH. NO. and Smallest Pipe Sizes First.
Example: 3" STD. L.R. 90° ELL
4" SCH. 160 L.R. 90° ELL
1½" EX-HVY L.R. 90° ELL

Fig. 11-5 Suggested order for items in "Bill of Material."

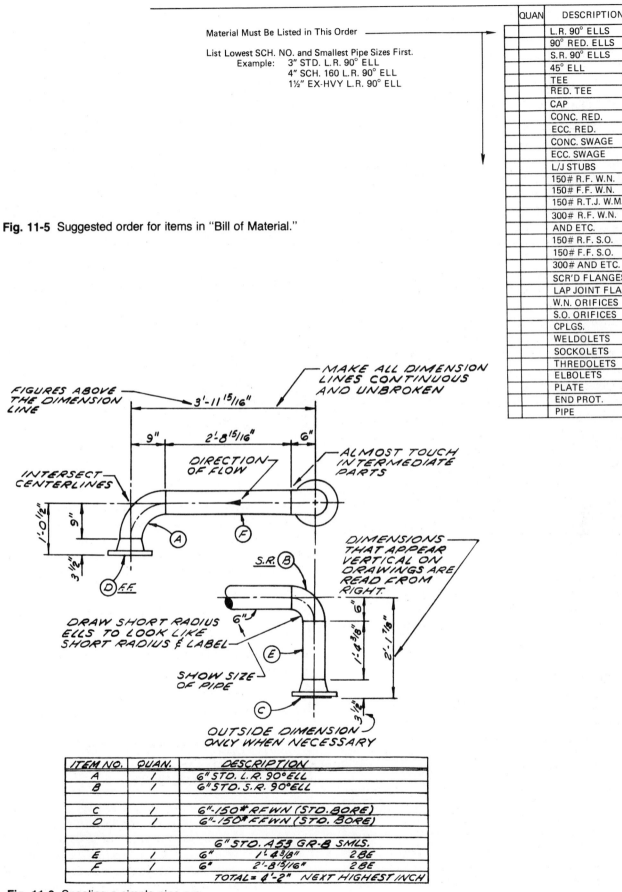

ITEM NO.	QUAN.	DESCRIPTION	
A	1	6" STD. L.R. 90° ELL	
B	1	6" STD. S.R. 90° ELL	
C	1	6"-150# RFWN (STD. BORE)	
D	1	6"-150# FFWN (STD. BORE)	
		6" STD. A53 GR-B SMLS.	
E	1	6"	1'-4 3/8"	2 BE
F	1	6"	2'-8 15/16"	2 BE
		TOTAL = 4'-2" NEXT HIGHEST INCH	

Fig. 11-6 Spooling a simple pipe run.

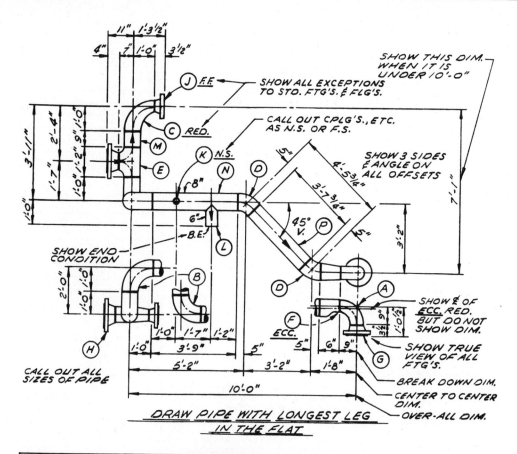

Fig. 11-7 Spooling several offsets.

ITEM NO.	QUAN.	DESCRIPTION		
A	1	6"STD. L.R. 90°ELL		
B	2	8"STD. L.R. 90°ELL		
C	1	8"x 6"STD. 90°RED. ELL.		
D	2	8"STD. 45°ELL		
E	1	8"STD. TEE		
F	1	8"x 6"STD. ECC. RED.		
G	1	6"-150# WNRF (STD. BORE)		
H	1	8"-150# WNRF (STD. BORE)		
J	1	6"-150# WNRF (STD. BORE)		
K	1	1"-3000# CPLG. SCR'D.		
		STD. A-53 GR-B SMLS.		
L	1	6"	0'-8 13/16"	1PE-1BE
M	1	8"	0'-9"	2BE
N	1	8"	3'-9"	2BE
P	1	8"	3'-7 3/4"	2BE
		TOTAL = 8'-2" NEXT HIGHEST INCH		

ITEM NO.	QUAN.	DESCRIPTION		
A	1	8"STD. L.R. 90°ELL		
B	1	8"STD. 45°ELL		
C	1	6"ON 8"STD. W-O-L		
D	1	6"-150# RFWN (STD. BORE)		
		STD. A53 GR-B SMLS.		
E	1	6"	0'-6"	2BE
F	1	8"	5'-7"	2BE
G	1	8"	4'-3"	2BE
		TOTAL: 9'-10"		

Fig. 11-8 Spooling a 45° angle.

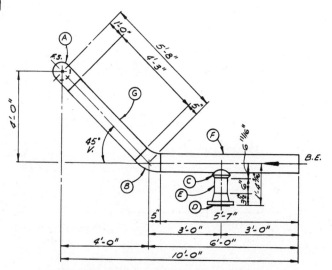

TYPICAL DIMENSIONING OF 45°ANGLE

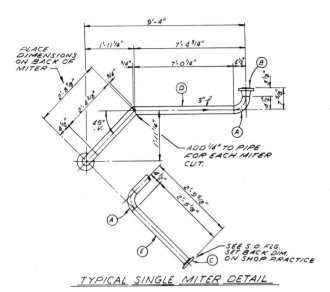

TYPICAL SINGLE MITER DETAIL

ITEM NO.	QUAN.	DESCRIPTION	
A	2	3" X-HVY L.R. 90° ELL	
B	1	3"-300# WNRF (X-HVY BORE)	
C	1	3"-300# SORF	
		X-HVY A53 GR-B SMLS.	
D	1	9'-4 7/8"	1 MITER END + 2BE
E	1	2'-4 3/4"	1PE x 1BE
		TOTAL = 11'-10"	

Fig. 11-9 Spooling a 45° miter.

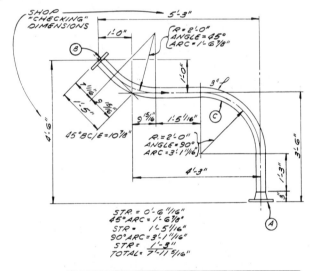

STR. = 0'-6 11/16"
45° ARC = 1'-6 7/8"
STR = 1'-5 11/16"
90° ARC = 3'-1 11/16"
STR. = 1'-3"
TOTAL = 7'-11 5/16"

ITEM NO.	QUAN.	DESCRIPTION	
A	1	3"-150# WNRF (STD. BORE)	
B	1	3"-150# SORF	
		STD. A53 GR-B SMLS.	
C	1	7'-11 5/16"	1BE-1PE

Fig. 11-10 Spooling a pipe bend section.

PIPE-FABRICATION METHODS

Pipe-fabrication methods are governed to a considerable extent by the rules of the *American Standard Code for Pressure Piping* (ANSI). Section 6 of the code, particularly, offers limits for design and rules for fabricating pipe and for qualifying welders. To

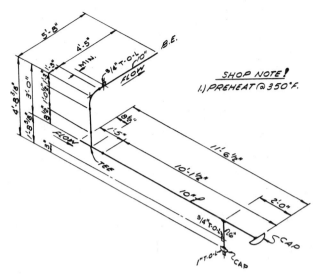

Fig. 11-11 Spooling a simple isometric run.

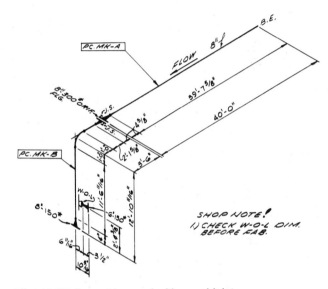

Fig. 11-12 Isometric spool with a weldolet.

supplement the various rules of the ANSI code, the Pipe Fabrication Institute (PFI) has prepared a series of standards for pipe fabrication. Additions to and revisions of these standards are being issued continually to pipe fabricators who are members of the association. The aims of these standards are:

A. To protect both purchaser and fabricator by providing consistent specifications and servicers.

B. To work toward standardizing fabricating methods.

C. To establish limits of responsiblity and to fix tolerances.

When better established or accepted, the rules will perhaps be comparable to the practices in shop fabrication of structural steel as defined by the American Institute of Steel Construction handbook.

EXERCISES

11-1. Match the terms and abbreviations:

1. swg	1. Bevel large end
2. B.L.E.	2. Large end plain
3. B.S.E.	3. Thread large end
4. B.B.E.	4. Bevel small end
5. B.E.P.	5. Swage nipple
6. L.E.P.	6. Sockolet
7. S.E.P.	7. Threadolet
8. T.B.E.	8. Thread both ends
9. T.L.E.	9. Small end plain
10. T.S.E	10. Bevel both ends
11. W.O.L.	11. Both ends plain
12. T.O.L.	12. Butt weld
13. S.O.L.	13. Field weld
14. E.O.L.	14. Flat face
15. S.W.	15. Bevel end
16. B.W.	16. Elbolet (state whether thread., B.E., or S.W.)
17. B.E.	17. Weldolet
18. P.E.	18. Thread small end
19. T.E.	19. Reducing
20. F.W.	20. Threaded end
21. F.F.	21. Plain end
22. red.	22. Socket weld
23. F.S.	23. Weld neck
24. N.S.	24. Raised face
25. W.N.	25. Far side
26. R.F.	26. Near side
27. S.O.	27. Jack screw
28. T.B.	28. Taper bore
29. orif.	29. Slip-on
30. J.S.	30. Mark
31. cplg.	31. Coupling (state whether thrd. or S.W.)
32. mK.	32. Orifice flange
33. thrd.	33. Threaded
34. ANSI	34. Specification
35. spec.	35. American National Standards Institute
36. std.	36. Standard

11-2. Write a brief report on pipe-fabrication methods and procedures.

11-3. Contact several companies and collect shop spool drawings from those companies.

11-4. Visit a local refinery, pump station, mechanical room or office building and inspect various types of valves, fittings, and sizes of pipe. Take pictures, and sketch the various spools in the sytem.

11-5. Select one of the isometric spools in this chapter and make a material list of fittings and pipe.

JOB ASSIGNMENTS

11-6. Redraw the pipe spool in Fig. 11-7 as shown.

11-7. Redraw the pipe spool in Fig. 1-6 as a single-line isometric (include bill of material).

11-8. Redraw the pipe spool in Fig. 11-7 as a single-line isometric (include bill of material).

11-9. Redraw Fig. 11-13 as a two-line orthographic (make out a bill of material for its fittings and pipe).

11-10. Change an isometric spool drawing to an orthographic single-line spool (see Fig. 11-13). Add bill of material.

11-11. Make a double-line orthographic spool drawing of line shown in Fig. 11-14 (include dimensions and bill of material).

11-12. Make a set of four spools (A, B, C, and D) for the line shown in Fig. 11-15. Use either orthographic or isometric spool technique.

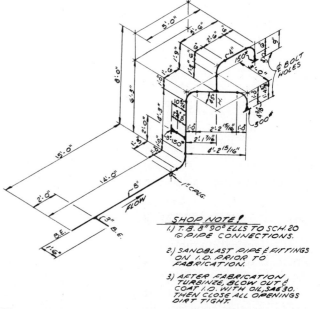

Fig. 11-13 Isometric spool of a rotated ell section.

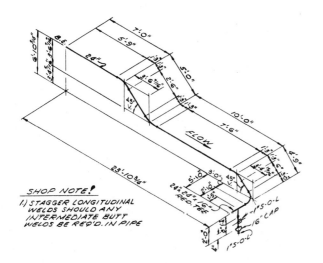

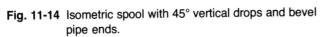

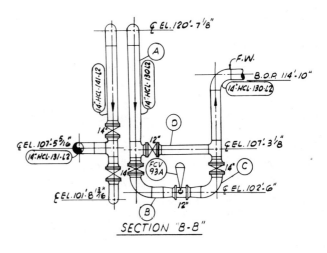

Fig. 11-14 Isometric spool with 45° vertical drops and bevel pipe ends.

Fig. 11-15 Spooling exercise.

Twelve

Specifications–Working Drawings and Design Practices

This chapter discusses briefly some of the basic considerations in writing specifications. It also discusses how drafting procedures must conform to certain criteria so that a set of plans is complete—i.e., so that the project can be completed from the information given.

A complete set of working drawings consists of specifications, plot plans, flow diagrams, plan and elevations, one-line isometrics, spools, structural details, electrical drawings, architectural drawings, vendor drawings, equipment details, miscellaneous details, and whatever else may be required to completely describe the pipe project. The pipe drafter should be familiar with all phases of the piping project and should be able to complete whatever pipe drawings are needed. The ability to do all phases of the work requires years of experience and training coupled with a willingness to tackle and complete whatever comes along.

In a typical refinery or pipe complex, several different types of fluids and gases must be transported. Each pipe system operates at a different temperature and pressure and so must be made from certain materials. If each situation is not handled individually, then some pipe will be overdesigned and some may be underdesigned. This in fact is what happened during the early years of piping design. Numerous faults such as explosions, leakages, and pressure drops caused many accidents and poor efficiency in plant operation.

Figure 12-1 is an example of a list of fluids and gases that should be prepared prior to outlining a complete set of specs. Word descriptions are usually written in sentence form at first, and are then abbreviated into a form such as in Fig. 12-2. There are many manufacturers for all components, but the specs must be written in a form that does not suggest any particular company.

Smaller companies often allow the drafter to do all phases of each job, while in larger companies each job is done by a separate department that specializes in that area. Whatever the situation, the drafter must begin with the simpler jobs and move to the more complex.

Spool-fabrication jobs at first seem complicated but certainly are at the lower end of the spectrum of difficulty. Although a thorough knowledge of fittings, valves, flanges, pipe, and equipment is required, one must go far beyond this level in order to be able to lay out and design a plant. A good pipe layout drafter must work from flow sheets and plot plans to derive a workable plant layout. The following pages briefly summarize what is required on each part of a set of working drawings. While studying this material, it would be beneficial to refer to the sample set of working drawings in the Appendix.

Fig. 12-1. Line contents.

Fluid designation	Service and fluid	Operating pressure, psig	temperature, °F
HC 800 Maximum	Hydrocarbon #1* (heater/crossovers)	250	800 Maximum
HB	Hydrocarbon #2*	250	800
HA	Hydrocarbon #3^	250	750 Maximum
H	Hydrocarbon #4*	100	600
S	Steam	275	410
SE	Steam, exhaust	40	290
SC	Steam, condensate	15	240
A	Air, utility	100	Ambient
AC	Air, instrument, dried	50	Ambient
F	Fuel, natural gas, dry	50	Ambient
C	Caustic	50	Ambient
IN	Instrument piping	Same as line services	
WF	Water, fire protection	100	Ambient
WC	Water, process cooling	75	85
W	Water, utility	75	Ambient
WS	Water, sanitary	40	Ambient

Fig. 12-2. Base line specifications.

Specification	Service	Flanges	Pipe	Valves	Reasons for Category
M	SE, SC, A, H, F, C	150 lb ASA	2″ and below: Sched. 80 ASTM A53 Gr A or ASTM A83 3″ to 10″: Sched. 40 ASTM A53 Gr A	Cast steel 150 lb flanged valves and forged steel screwed valves	These services fall into 150 lb ASA class and require all steel materials, no cast iron (see Code). ASTM A53 pipe is satisfactory for services below 900°F. A83 is indicated as an alternate in case A53 is not available in small sizes.
N	S, HA, HB, HC	300 lb ASA	2″ and below: Sched. 80 ASTM A106 Gr A 3″ to 10″: Sched. 40 ASTM A106 Gr A 2″ and larger: 0.375″ wall (0.375″ wall commonly available in these large sizes)	Cast steel 300 lb flanged valves and forged steel screwed valves	Another "all steel" spec. requiring 300 lb flange rating. ASTM A106 Gr A is used here as a safety factor since operating temperatures of 800° are rather close to the limit of 900°F. for ASTM A53.
P	WF, W, WC, WS	150 lb ASA	Same as in N for above ground. Use cast iron, below ground and ASTM A120 galvanized for WS (drinking water)	Cast iron flanged valves and brass screwed valves	Cast iron can be used for valves in water service. In itself this economy justifies a separate specification for water. In addition, underground pipe is cast iron, and galvanized pipe must also be specified. 150 lb ASA steel flanges are used so that piping may be welded.
Q	AC	125 lb ASA Cast iron	Same as above ground pipe for N	Same as in P	These lines will be small and the cheaper cast iron screwed flanges will be used. Also only steel pipe is used for air. These two facts along with other minor details justify a separate spec.
Z	IN	Same as line service	Same as line service-Copper tubing for air transmission	Same as line service	Instrument lines require special notes and the air transmission lines are tubing thus justifying a separate spec.

SPECIFICATIONS

Specifications are one of the most important parts of a set of working drawings. Specifications are the guidelines to which the entire plant must conform. Without specifications, sizes, thicknesses, physical properties, etc., would not at times be adequate to meet the demands of the situation. General specifications with all companies are similar in that the information is usually given in a definite sequence. Some specifications are quite involved. Others may be simple and to the point. Prior to beginning a job, the drafter must study very thoroughly the "rules of the game," so to speak. In the Appendix along with the set of plans there is a sample set of specifications given in the same sequence as a typical set of specs for a pipe project. Major topics are numbered; for example, topic 3 is "Design." Subtopics under "Design" are indicated, such as 3.1, which is "General." The paragraphs under it, such as 3.1.1, refer to the condition of the pipe and the material to be used.

Although this set of specs is typical, it by no means is complete, nor could it be used for all projects. Specifications should be read and studied thoroughly prior to starting a job.

WORKING DRAWINGS

As was discussed in the introduction of this chapter, a set of working drawings consists of specifications and all the other actual drawings required to complete a job or project. An elaborate explanation of each would be quite lengthy and involved. The following discussion deals very briefly with each type of drawings. A sample set of plans is included in the Appendix and should be studied thoroughly.

The following is a listing of the types of drawings:

- Flow diagram
- Instrumentation
- Plot plan
- Vessels and equipment
- Foundation plan
- Structural design
- Pipe plans, elevations, and sections
- One-line isometrics
- Spools
- Electrical
- Architectural
- Vendor drawings and standards

Flow Sheet: A flow sheet (Fig. 12-3) is developed prior to most other drawings. It should be a schematic (no-scale) drawing, used to derive a plan, ele-

vation, and section to scale for every area of the plant. Coordination of design criteria, plot plans, and flow sheet data gets the piping in a logical sequence and workable arrangement. Flow sheets are generally of three types: *process, mechanical,* and *utility.*

Flow sheets show instrumentation lines that control the *flow, pressure,* and *temperature* of the entire facility. Beginning drafters and even experienced designers often have difficulty in working from flow sheets and perhaps even more difficulty laying out a flow sheet. The flow sheet is simply a schematic representation of a chemical engineer's idea that is on its way to becoming a reality. Study the mechanical flow sheet in the Appendix.

Instrumentation: Instrumentation drawings show the location of all control panels and, in turn, the path of all railways, instruments, and details. *Instruments* are the control mechanisms used to open and close valves throughout a system. Instrumentation design requires a thorough knowledge and understanding of all phases of process, mechanical, and utility flow systems.

Symbols and terminology for flow and instrumentation are very similar, even though their functions are quite different. Since there is such a similarity, the two will be discussed in conjunction.

INSTRUMENTATION TECHNOLOGY

- Bubbles—circle used to call out a type of control device
- Operating pressure—normal operating pressure
- Operating temperature—normal operating temperature
- Design pressure—overdesign in pressure rating to allow for extremes
- Design temperature—overdesign in temperature rating to allow for extremes
- Process flow—basic overall flow pattern
- Mechanical flow—pattern
- Vertical coordinates—control measurements that run up and down as height measurements
- Plan coordinates—north-south and east-west control dimensions
- Line number—identification of each line of pipe
- Gauge glass—sight glass for determining liquid level
- Battery limit—major outside boundaries of a plant
- Plant operator—individual in charge of a plant facility

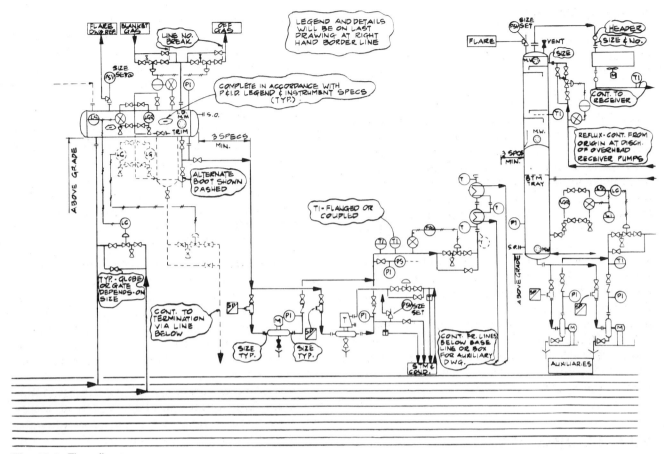

Fig. 12-3 Flow diagram.

- Control station—central control location for all liquid and gas flow
- Thermowell—device used to sense temperature
- Plot plan—general layout of a plant
- Utility flow—flow diagram for all light, water, air, and other utility items

Flow and Instrument Symbols: There are four basic groups of instruments, namely, *flow*, *level*, *pressure*, and *temperature*. An instrument may be locally mounted, or it may be mounted on board. Board-mounted instruments send signals to an instrument panel control board. The symbol for a locally mounted instrument is an open circle. A board-mounted instrument is symbolized by a circle with a horizontal line.

Instruments of all four types may perform one or all the following functions:

- Control
- Record
- Indicate
- Alarm

Figure 12-4 illustrates various instrument symbols used in pipe drafting. The student should become familiar with these and learn to recognize them on pipe blueprints and flow sheets. In Fig. 12-5 valve lines and miscellaneous abbreviations used in instrument drawings are shown fro the student to study and use for reference.

The symbols shown in Figs. 12-4 and 12-5 (pages 144 and 145) are combined in various forms to depict several of the more common applications of flow and instrument symbols. The flow diagram symbol is shown as well as how that arrangement would look on a complete pipe drawing. Study Figs. 12-6 to 12-31 (pages 146 to 152) very carefully and notice each symbol's particular application. The flow symbols are broken down into four categories: temperature, pressure, flow, and level. Actual systems are combinations of these four types of flow.

Figs. 12-6 to 12-11—temperature
Figs. 12-12 to 12-19—pressure
Figs. 12-20 to 12-23—flow
Figs. 12-24 to 12-29—level recorder
Figs. 12-30 and 12-31—flow sheet drawings

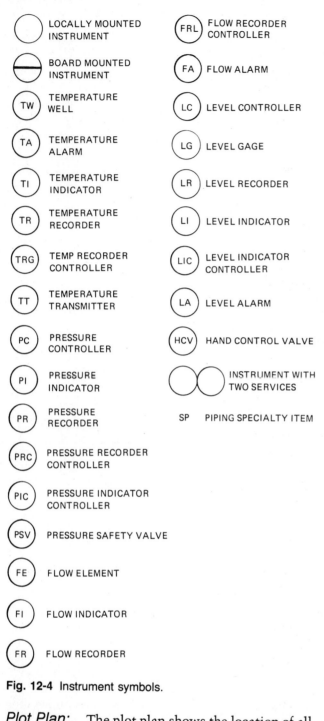

◯	LOCALLY MOUNTED INSTRUMENT
⊖	BOARD MOUNTED INSTRUMENT
TW	TEMPERATURE WELL
TA	TEMPERATURE ALARM
TI	TEMPERATURE INDICATOR
TR	TEMPERATURE RECORDER
TRG	TEMP RECORDER CONTROLLER
TT	TEMPERATURE TRANSMITTER
PC	PRESSURE CONTROLLER
PI	PRESSURE INDICATOR
PR	PRESSURE RECORDER
PRC	PRESSURE RECORDER CONTROLLER
PIC	PRESSURE INDICATOR CONTROLLER
PSV	PRESSURE SAFETY VALVE
FE	FLOW ELEMENT
FI	FLOW INDICATOR
FR	FLOW RECORDER

FRL	FLOW RECORDER CONTROLLER
FA	FLOW ALARM
LC	LEVEL CONTROLLER
LG	LEVEL GAGE
LR	LEVEL RECORDER
LI	LEVEL INDICATOR
LIC	LEVEL INDICATOR CONTROLLER
LA	LEVEL ALARM
HCV	HAND CONTROL VALVE
◯◯	INSTRUMENT WITH TWO SERVICES
SP	PIPING SPECIALTY ITEM

Fig. 12-4 Instrument symbols.

Plot Plan: The plot plan shows the location of all major equipment on the plant site. Roads, walkways, maintenance accesses, buildings, and foundations are located according to specification. The plot plan is used as a key for the design of the total plant facility, with sheet identifications and grid coordinates used for cross-reference. The plot plan is one of the first sheets to be developed and is constantly referred to for basic information.

Vessels and Equipment: Preliminary drawings of vessels are used during the early stages of plant design. Later, as detailed drawings of each vessel and piece of equipment are prepared, the final dimensions of nozzle projections, locations, or other factors may cause the pipe designer to alter the plan. Vessels are detailed by custom fabrication shops that specialize in this type of design.

Foundation Drawing: A foundation drawing shows dimensions and outlines of a foundation, location and size of reinforcing steel, and bending details for steel. The piping designer must work closely with the foundation designer since underground water systems and sewers require large piping which can interfere with foundations. Thus underground piping, underground electrical conduit, and foundations are developed simultaneously. Often all three are ultimately shown on the same drawings or associated drawings, so that underground work may proceed simultaneously.

Structural Drawing: Structural steel and concrete are used not only in the architecture but also to support tanks, equipment, pipe, and the liquid flowing through it in many situations. Also, structures must be designed to support loads of wind, water, and vibration. Engineering is critical. Structural design and detail are a very important supportive type of pipe drafting. There should be a very close relationship between the pipe layout, structural design, and drafting departments.

Plans, Elevations, and Sections: The *plan view* of a pipe plant means a top view of definite area of the total coordinate plot plan. Plan views also can be taken from several elevations within the same coordinate area. A plan view of several levels combined would be too complicated and busy to work from. Plan views could, for example, be at ground level, arbitrarily designated as 100 ft 0 in, and plans could be taken at levels of 108 ft 0 in, 112 ft 0 in, etc. The plans and elevations in the Appendix are not a complete set but are only of three coordinate areas, D1, D2, and D3, which consist of three plan views, 506–50–126, 506–50–134, and 506–50–142. Since this plant is rather simple and of a low profile, only one plan level is needed. Numerous sections and elevations are taken from the plan view to clarify the many hookups and pipe runs which cannot be shown on just a plan view. Sections such as section B of sheet 506–50–234 give a great deal of information not seen from just a plan view. Study the plans and sections in the Appendix for a more thorough understanding of their part in the total set of working drawings.

Fig. 12-5 Valve and line symbols.

One-line Isometrics: As has been stated in previous chapters, pipe is drawn as plans and elevations, but individual pipelines must be drawn on an individual basis and dimensioned as individual lines rather than as a total assembled pipe unit. One-line isometrics are used to show all the pipe components and center-to-center dimensions. The line designation of each line describes its size, contents, number, and specifications. Each line has its own number. See set of plans in the Appendix. This sheet is a drawing of line 14″–HCL–141–A–L2. A standard bill of material and titleblock sheet are used with

listings of all types of flanges, fittings, valves, pipe, etc. Time is saved by just filling in the quantities and sizes of each line component. Items listed for this line isometric are tabulated by material takeoff persons and used to order all the warehouse stock items that will be needed later to fabricate individual pipe spools. One-line isometrics are not absolutely necessary and some jobs bypass this phase, but doing so slows down the spooling process a great deal. Line isometrics cut down on many unnecessary mistakes that are often made in spooling directly from plans with elevations. Line isometrics

145

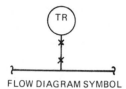

FLOW DIAGRAM SYMBOL

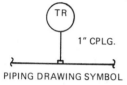

1" CPLG.

PIPING DRAWING SYMBOL

Fig. 12-6 A temperature recorder (TR) is a recorder that provides a permanent, continuous record of equipment or pipe temperature.

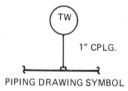

FLOW DIAGRAM SYMBOL

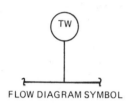

1" CPLG.

PIPING DRAWING SYMBOL

Fig. 12-7 A temperature well (TW) is a unit used for the protection of the temperature instrument bulb.

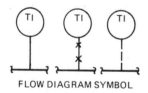

FLOW DIAGRAM SYMBOL

1" CPLG.

Fig. 12-8 The temperature indicator (TI) can be of three types:
(1) a local-mounted dial thermometer,
(2) a remote-mounted dial with a cap tube to the measurement point, or
(3) an electric thermocouple with a remote-mounted indicator.

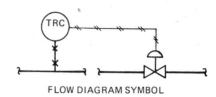

FLOW DIAGRAM SYMBOL

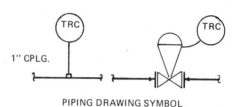

1" CPLG.

PIPING DRAWING SYMBOL

Fig. 12-9 The temperature recording controller (TRC) simultaneously records and regulates the line or vessel temperature.

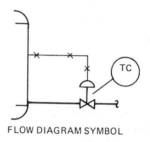

FLOW DIAGRAM SYMBOL

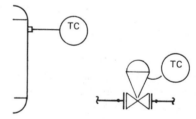

PIPING DRAWING SYMBOL

Fig. 12-10 The temperature controller (TC) is a control valve with a direct capillary connection to the point of measurement.

are not to scale but must be properly dimensioned and accurately billed.

Spool Drawings: Spooling and fabrication have already been discussed quite thoroughly. Spools are the actual pipe assembly pieces that are used to connect equipment at the job site. In the event engineering and plant layout are in error, spools that do not fit will be the result. Individual runs of pipe, or *spools* as they are commonly called, bear the same

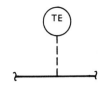

FLOW DIAGRAM SYMBOL

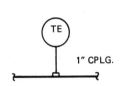

1" CPLG.

PIPING DRAWING SYMBOL

Fig. 12-11 Temperature elements (TE) are thermocouples with no connections to the instrument.

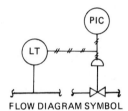

FLOW DIAGRAM SYMBOL

Fig. 12-12 The pressure indicator controller (PIC) is a pressure control valve with either an indicating type transmitter or a remote-mounted pressure indicator.

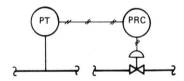

FLOW DIAGRAM SYMBOL

Fig. 12-13 Pressure recording controller (PRC) are similar to the pressure recorders. They have an added pneumatic signal to the control valve.

FLOW DIAGRAM SYMBOL

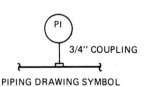

3/4" COUPLING

PIPING DRAWING SYMBOL

Fig. 12-14 The pressure indicator (PI) is a dial instrument that indicates line or equipment pressure.

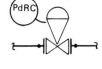

PIPING DRAWING SYMBOL

Fig. 12-15 A pressure differential recording controller (PdRC) controls the pressure differential between two vessels or pipe lines.

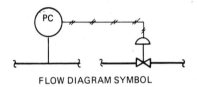

FLOW DIAGRAM SYMBOL

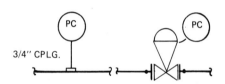

3/4" CPLG.

PIPING DRAWING SYMBOL

Fig. 12-16 The pressure control (PC) regulates the pipe or vessel pressure. It is connected to the measurement point with a pneumatic transmitter.

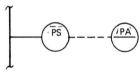

FLOW DIAGRAM SYMBOL

3/4" CPLG.

PIPING DRAWING SYMBOL

Fig. 12-17 Pressure alarms (PA) are pressure switches that are attached, by a coupling or a flange, to the pipe.

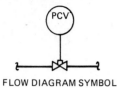

FLOW DIAGRAM SYMBOL

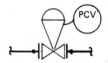

PIPING DRAWING SYMBOL

Fig. 12-18 Pressure control valves (PCV) are self-contained valves that regulate pressure.

FLOW DIAGRAM SYMBOL

PIPING DRAWING SYMBOL

Fig. 12-19 The pressure safety valve (PSV) is an automatic pressure-relieving device that is triggered by pressure upstream of the valve.

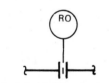

FLOW DIAGRAM SYMBOL

PIPING DRAWING SYMBOL

Fig. 12-20 A flow restriction orifice (RO) is a standard for special flanges or unions that hold plates with a small hole, causing constant flow.

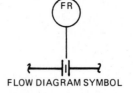

FLOW DIAGRAM SYMBOL

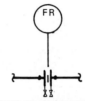

PIPING DRAWING SYMBOL

Fig. 12-21 The flow recorder (FR) makes a permanent record of the flow measurement.

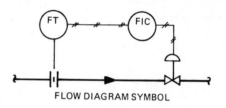

FLOW DIAGRAM SYMBOL

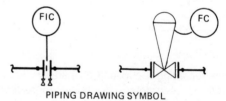

PIPING DRAWING SYMBOL

Fig. 12-22 A flow indicator controller (FIC) is a flow indicator and a control valve.

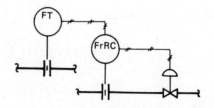

FLOW FIAGRAM SYMBOL

Fig. 12-23 A flow ratio recording controller (FrRC) is a control valve and a recording instrument that records and controls the flow ratio of the main line.

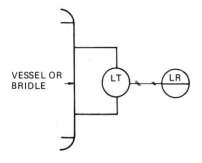

FLOW DIAGRAM SYMBOL

Fig. 12-24 Level recorders (LR) make permanent records of the liquid level in vessels. They use pneumatic signals from displacement-type transmitters on the vessels.

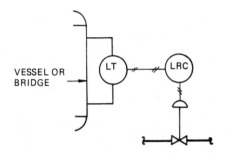

FLOW DIAGRAM SYMBOL

Fig. 12-25 Level recorder controllers (LRC) are the same as level recorders. However, they use a pneumatic signal to the control valve as well as a recorder.

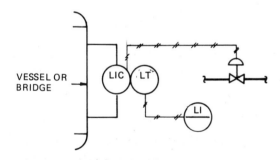

FLOW DIAGRAM SYMBOL

Fig. 12-26 A level indicating controller (LIC) is a simultaneous control and indication of liquid level.

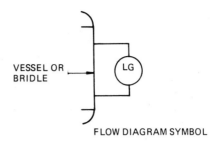

FLOW DIAGRAM SYMBOL

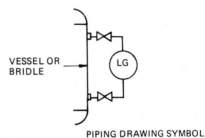

PIPING DRAWING SYMBOL

Fig. 12-27 A level glass (LG) is a direct reading device that is connected at the low and high points of the level variation. The liquid in the tank seeks its own level and can be observed through a transparent glass in the instrument.

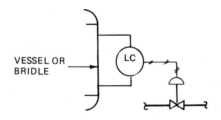

FLOW DIAGRAM SYMBOL

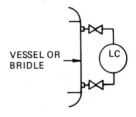

PIPING DRAWING SYMBOL

Fig. 12-28 The level controller (LC) is a blind instrument that regulates the vessel liquid by a pneumatic signal to the control valve.

line number as the entire line from which they came, such as 14″–HCL–141–A–L2. Spools are individualized by adding dash numbers to the main line designation number, such as –1, –2, –3.

The size of a spool is limited by the fabricator's capabilities for shipping it to its location. The standard overall space is 40′ × 8′ × 8′. A spool cannot be longer, wider, or taller than this.

Spools are usually fabricated in a shop for sizes 3″ or larger and in the field for smaller sizes.

As has been covered earlier, spools are instructions to the welders and erection crew and should include drawings, bills of material, notes, and information on how they are to be assembled. Spools can

149

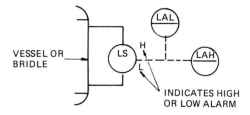

FLOW DIAGRAM SYMBOL

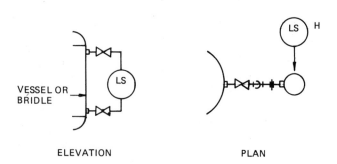

ELEVATION   PLAN

PIPING DRAWING SYMBOL

Fig. 12-29 The level alarm (LA) indicates high- or low-liquid level by an electric signal.

be very simple, but some are very complex and require a lot of experience.

Spooling is an excellent place for the trainee pipe drafter to begin, in that the drawings deal with very basic components and are small enough for the drafter to complete in a reasonable amount of time.

Electrical Drawings: Electrical drawings are normally associated with control of flow as in instrumentation, but the electrical drawings used for utilities are essential and so are considered a very important part of the total working drawing project. They can become a very specialized supportive area.

Electrical drawings are *schematic*—nondimensioned drawings used to locate conduit, lighting fixtures, circuits, and numerous other electrical information. Since electrical conduit is more easily arranged and generally smaller than pipe, it is common practice to install electrical features around pipe runs. Some electrical conduit, however, is 3° or larger and must be planned just as carefully as process pipe.

Architectural Drawings: Architectural drawings seem an unlikely subject for pipe drafters, but numerous shops, offices, laboratories, and test areas must be detailed. Outside architects are usually not familiar with pipe, and as a result their designs often create some functional problems. Architecture is a separate area but must be coordinated closely with approved piping methods.

Vendor Drawings: Vendors are basically manufacturers that produce items that require little engineering but can be selected to fit a given situation. Vendor drawings include standard dimensions and details of equipment that are supplied and must be installed in a pipe system according to specifications.

Other types of drawings, such as of equipment details, pumps, compressors, heat exchangers, heaters, separators, and condensors, must be prepared by companies that specialize in detailing and fabricating such component units. Although often thought of as a separate area, such drawings must be fully coordinated to ensure consistency of drawings.

Conclusion: There are perhaps other types of supportive drawings, but we have covered the most common. All are vital to the total plant operation. Experience will enable the drafter to work with all types of drawings that are supportive to the piping process.

BASIC DESIGN PRACTICES

Chapters 1 to 11 deal primarily with the fundamental concepts of pipe, equipment, valves, flanges, and many of the various drafting procedures necessary to properly describe a piping system. The previous section of Chap. 12 is a brief introduction to some basic design factors, such as space requirements, safety, and code requirements. This section of the chapter is not an attempt at completely covering all facets of design. To do so would require several books. After completion of this text material, students should continue their study in specific areas of design related to their interest or job classification.

Basic Design Criteria: The piping flow engineer on a project is a key person responsible for the coordination of all piping drawings. Specifications, flow diagrams, nomenclature, and equipment locations are used to determine the other engineer working drawings. This basic information is expanded to various specialized departments capable of producing quality work in their specific areas. There must be a great deal of coordination among these specific areas. The flow diagram is the basic source of information and all design must adhere to its guidelines. Flow lines are broken up into individual lines, and each line must be properly sized, classified, and recorded. The flow sheet and line nomenclature are like a road map giving direction of flow in a sequence of operation. Most mistakes are made from simple, old, established, design errors. Basic mistakes made during the construction stages can create endless problems in time and material.

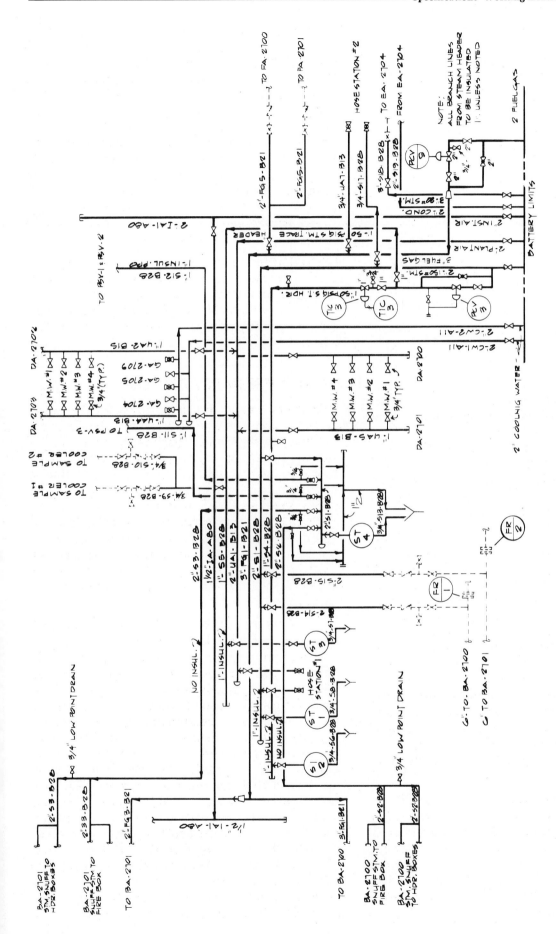

Fig. 12-30 Flow sheet diagram.

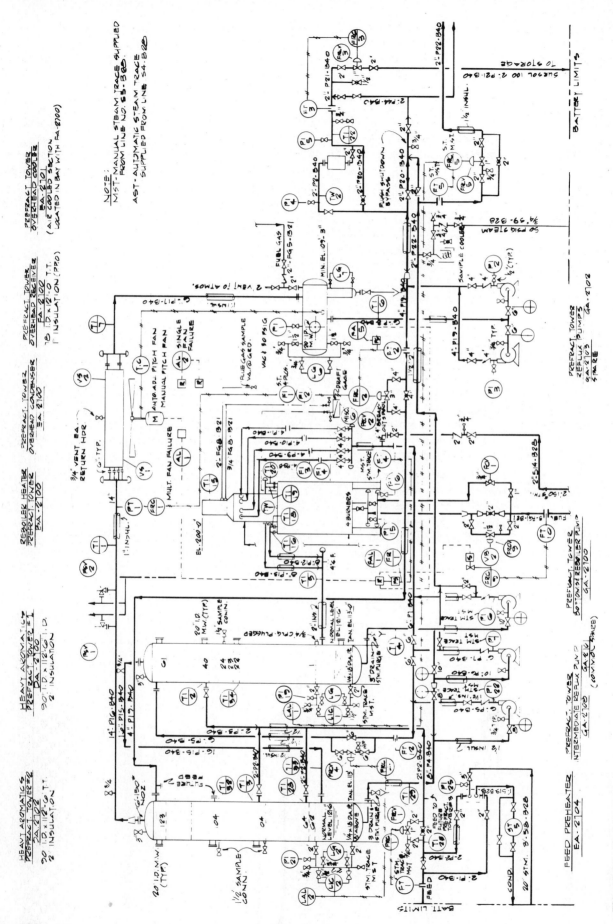

Fig. 12-31 Flow sheet diagram.

Design Problems and Errors: Most facilities have problems and mistakes that were caused by designers and engineers who did not conform to established criteria, had low employee professionalism, or lacked knowledge of the particular work at hand. Whatever the problem, there is a solution that best suits the situation. Experience and patience in arriving at a solution based on all the data at hand are the scientific approach to the problem. Described below are some specific design problems as they occur in various stages of plant design.

Design Problems in Working Drawings:

Spacing of equipment, such as vessels, tanks, and pipe in order to make best use of usable space is one of the most basic design problems. *Location of the north arrow* so that all sheets conform to a systematic orientation process helps clarify coordinate dimensions and elevations. *Match lines as boundaries* for specified portions of the plot are a source of mismatching of lines from one sheet to another, and from one elevation to another.

Thermal contraction and expansion are a serious design problem, especially in areas where there are extremes of hot and cold temperature. Somewhat related to thermal expansion is the *variation of settling* among different towers and equipment.

Structural interferences create problems that are often not noted until the erection stages, and by that time the cost to change is excessive.

General Drafting Factors:

Improperly designed pipe systems add thousands of dollars to the erection, operation, and maintenance of a plant. Simple mistakes cause loss of time and money. Errors are generally due to not keeping up with new processes or failure to adhere to established standards and procedures. Design criteria aid the drafter in coming up with a workable plan. Each company has its own guidelines, some of which must conform to generally allied practices. Design factors can be categorized into several major areas ranging from basic drafting guidelines to major equipment location criteria.

Some very basic design and drafting guidelines for working drawings are listed below:

- Pipe that is standard and readily available should be used when possible.
- Vent and drain pipes at high and low points.
- Do not run pipe under foundations.
- Provide places to disassemble or remove equipment.
- Pipe should be positioned to provide adequate clearance for installation and maintenance.

- Make sure valves are within reach for operation.
- Provide room for expansion joints and loops to facilitate contraction and expansion due to hot and cold conditions.
- Provide adequate access to valves; consider function and frequency of use.
- Do not point valve handles into traveled passageways.
- Most valves operate better if handles are up or horizontal rather than in a down position.
- Select correct valve to fit design situation.
- Some valves need to be removed and replaced periodically; provide adequate space to do so.
- Locate control station near equipment it serves.
- Route pipe in a system on the basis of its function rather than the supports that are available.
- All pipelines should give direction of flow and should use appropraite line designation callout above line.
- Change line numbers at tanks, vessels, etc. But do not change them at smaller devices, such as valves.
- Plan view drawings should have match line numbers for adjoining plan views.
- Plan elevations and sections are usually drawn at a ⅜ in = 1 ft 0 in scale.
- Pipelines that need to be insulated should be so indicated on drawings.
- Companies use single-line as well as double-line pipe symbols as the need arises.
- Equipment is usually drawn schematically and is drawn lighter than the pipelines so that the pipe stands out.
- All pipe symbols should conform to ANSI standards.
- Buildings and structures should be shown light and in the background.
- North arrow (true or plant) should be shown on all plan view drawings pointing up or to the right side.
- Draw valves schematically but show dimensions such as face-to-face and handle height to scale.
- All weld symbols should conform to standards of the American Welding Society.
- All dimensions should be given in feet and inches, as 25' 3½".
- Instrumentation lines that cross should break, and all the breaks should be in one direction.
- Instrumentation lines should be approximately ⅜ in apart on flow sheets.

- Steam traps should be shown on drawings by symbol and number and if necessary detailed on a schedule drawing.
- Pumps should be provided with a check valve in the discharge line upstream from the block valve in order to prevent backflow.
- Pumps should be arranged with a minimum of 2 ft 6 in clearance between units and 5 ft 0 in aisle space. Clearance should be provided for part removal.
- Compressors should be arranged to avoid cyclic vibration.

Conclusion: The preceding piping guidelines are only a few of the many standards used to aid the drafter in making decisions. In an actual drafting situation the drafter would have numerous other guidelines, books, and standards to use in conjunction with a typical pipe design. To cover the subject of design would involve a complete and separate text in itself.

EXERCISES

12-1. Define the following terms briefly:

1. Specification
2. Working drawing
3. Line numbering procedure
4. Pipeline service
5. Drip leg
6. Strainer
7. Steam trap
8. Expansion loop
9. Thermowell
10. Bubble
11. Process flow
12. Gauge glass
13. Utility flow
14. Foundation
15. Spool drawing
16. Vendor drawing
17. Pipe model
18. Turbine
19. Compressor

12-2. Match the abbreviations to the following terms:

1. Fuel gas   **a.** BOP
2. Air   **b.** I.A.
3. Water   **c.** HCL
4. Hydrocarbon liquid   **d.** HCV
5. Diesel fuel   **e.** B.G.
6. American Society of Tubular manufacturers
7. American National Standards Institute
8. Hydrocarbon vapor   **f.** ANSI   **k.** SPEC
9. Instrument air   **g.** ASIM   **l.** F.W.
10. Blanket gas   **h.** D.F.   **m.** FG
11. Bottom of pipe   **i.** H_2O **n.** A
12. Field weld   **j.** PE
13. Plain end
14. Specification

12-3. Review Questions

1. List six types of working drawings.
2. Describe the purpose of specifications
3. Name three types of flow sheets.
4. What are the four groups of instruments used on instrumentation drawings?
5. Write a brief explanation of how the coordinate system of locations is used on plot plans.
6. How is a one-line drawing different from a spool drawing?
7. Write a sample line description and identify the components.
8. What is a vendor drawing?
9. What is the function of the ANSI?
10. What are some problems in good design procedure?
11. What scale should plans and elevations be drawn to?
12. What is the scale of flow sheets?
13. Describe briefly what battery limit is.
14. Why is it necessary for each pipeline to have its own specification criteria?

Appendix A

Sample Sets of Plans

SPECIFICATIONS
Refinery Process Piping
150# Carbon Steel R.F. Flanges

ITEM	LIMITS	GENERAL DESCRIPTION
Pipe	1½" down	Schedule 80 SMLS. carbon steel
	10" down	Schedule 40 SMLS. carbon steel
	24" down	.375 SMLS. carbon steel
Gate valve	2" down	600# FS standard weight Vogt SW-12111
	24" down	150# CS flanged RF crane #47X
Globe valve	2" down	600# FS standard weight Vogt SW-12141
	8" down	150# CS flanged RF crane #143X
Check valve	2" down	600# FS standard weight Edwards #838Y
	12" down	150# CS flanged RF crane #147X
Plug valve	1½" down	300# CS screwed Nord. #2024
	2" down	150# CS screwed Nord. #1924
	4" down	150# CS flanged RF Nord. #1925
	6" down	150# CS flanged RF Nord. #1945
	12" down	150# CS flanged RF Nord. #1949
	16" down	150# CS flanged RF Nord. #4169
Flanges	1½" down	150# FS standard weight RF
	24" down	150# FSWN RF bored to match pipe
	24" down	150# FS blind RF
Fittings, ells tees, etc.	1½" down	2000# FS standard weight
	24" down	Butt weld same schedule or wall thickness as pipe
Plug	2" down	2000# FS screened hex head
Coupling	1" down	2000# FS standard weight full (use screened coupling @)
	2" down	300# FS standard weight full (H.P. vents)
Union	2" down	3000# FS GJ standard weight w/integral seats
Gaskets	24" down	1/16" thick asbestos ring durabla
Bolts	24" down	Alloy steel stud bolts stamp B37 per A193
		Nuts per A194 stamp 2H

SPECIFICATIONS
Refinery Process Piping
300# Carbon Steel R.F. Flanges

ITEM	LIMITS	GENERAL DESCRIPTION
Pipe	1½" down	Schedule 80 SMLS. carbon steel
	10" down	Schedule 40 SMLS. carbon steel
	16" down	.375 wall SMLS. carbon steel
	20" down	.500 wall SMLS. carbon steel
	24" down	.562 wall SMLS. carbon steel
Gate valve	2" down	600# FS standard weight Vogt #SW-12111
	24" down	300# CS flanged RF crane #33X
Globe valve	2" down	600# FS standard weight Vogt SW-12141
	8" down	300# CS flanged RF crane #15IX
Check valve	2" down	600# FS standard weight Edwards #838Y
	12" down	300# CS flanged RF crane #159X
Wafer check valve	24" down	300# Steel, Mission Duo-Chek, 30 SMF (Buna-N seats) 30 SPF (steel seats) or equal for 250°F and below, use Buna N-to-metal seat. Above 250°F use metal-to-metal seat and specify Inconel "X" spring.
Plug valve	½" down	2000# forged steel, screwed, Nordstrom #2222 or equal
	2" down	300# CS screwed Nordstrom #2024 or equal
	4" down	300# CS flanged Nordstrom #2045 or equal
	24" down	300# CS flanged worn gear operated, Nordstrom #4289 or equal
Flanges	1½" down	300# FS standard weight RF
	24" down	300# FS WN RF bored to match pipe
	24" down	300# FS blind RF
Fittings, ells tees, etc.	1½" down	3000# FS standard weight
	24" down	Butt weld same schedule or wall thickness as pipe
Plug	2" down	2000# FS screwed hex head (H.P. vents only)
Coupling	1" down	6000# FS standard weight full (use screwed coupling @)
	2" down	3000# FS standard weight full (H.P. vents)
Union	2" down	3000# FS GJ standard weight w/integral seats
Gaskets	24" down	1/16" thick asbestos ring durabla
Bolts	24" down	Alloy steel stud bolts stamp B7 per A193
		Nuts per A194 stamp 2H

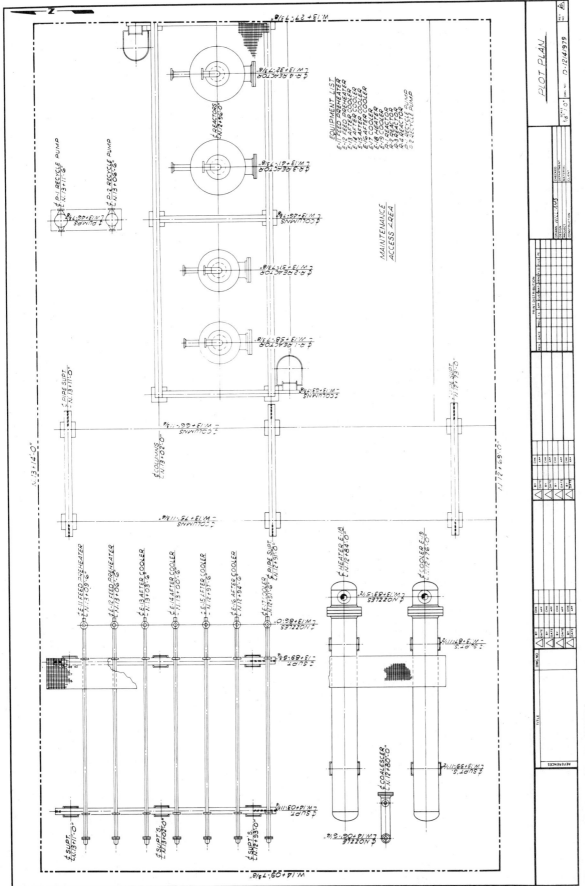

Fig. A-1 Plot Plan (D-12141979).

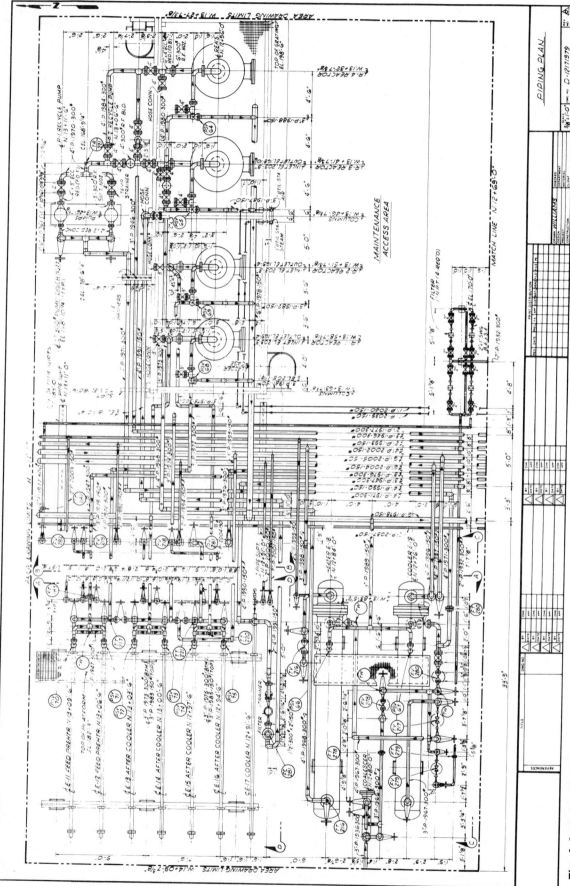

Fig. A-2 Piping Plan (D-1271979).

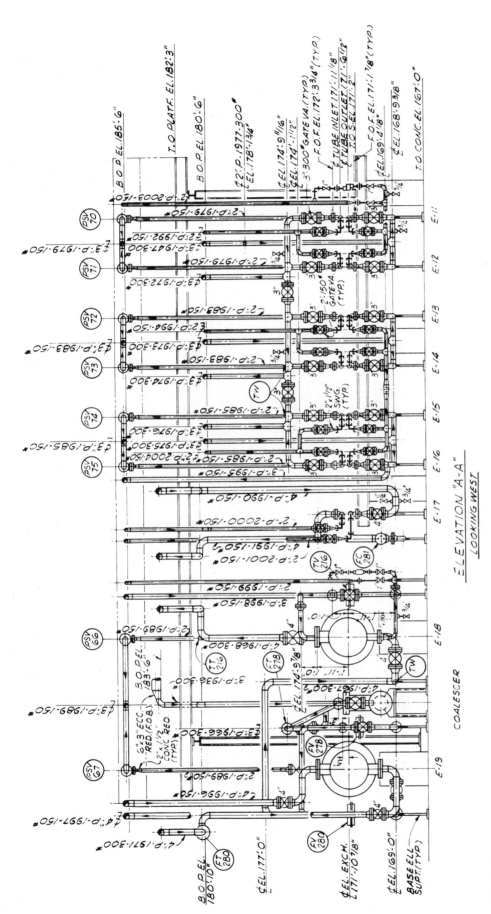

Fig. A-3 Elevation AA

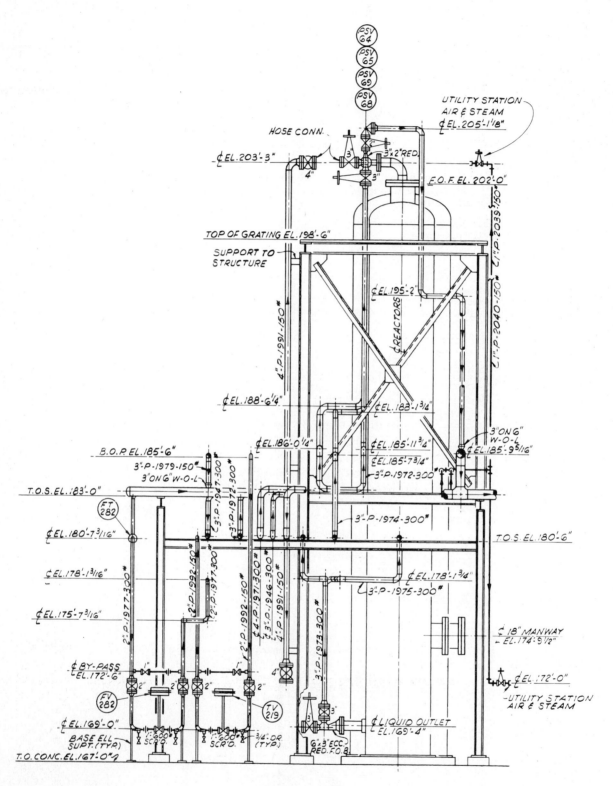

ELEVATION "B-B"
LOOKING EAST

Fig. A-4 Elevation BB

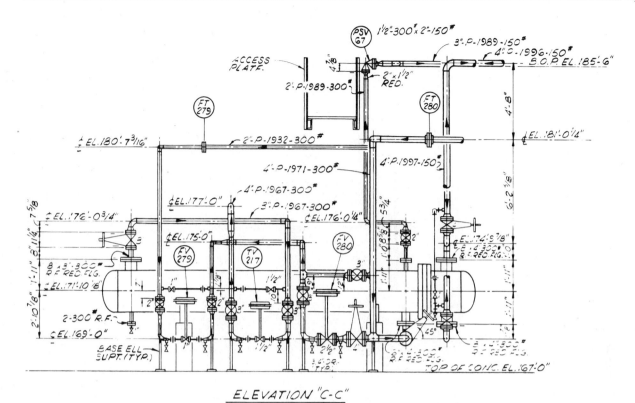

ELEVATION "C-C"
LOOKING NORTH

Fig. A-5 Elevation CC

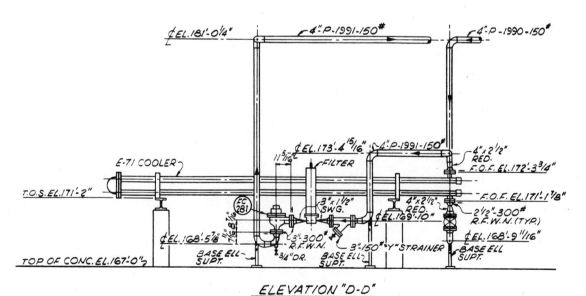

ELEVATION "D-D"
LOOKING NORTH

Fig. A-6 Elevation DD

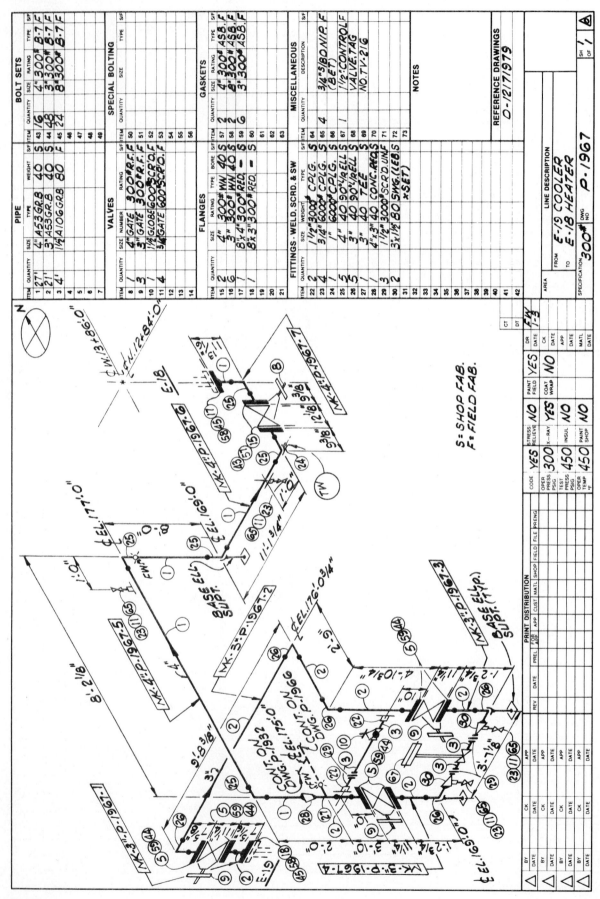

Fig. A-7 Line isometric drawing (P-1967)

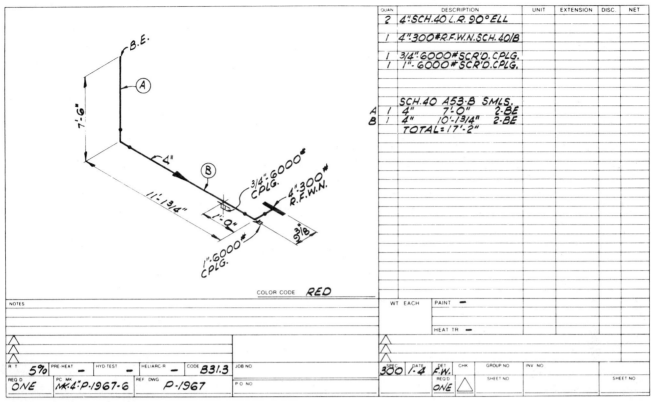

QUAN	DESCRIPTION	UNIT	EXTENSION	DISC.	NET
2	4"SCH.40 L.R. 90°ELL				
1	4"300#R.F.W.N. SCH.40/B				
1	3/4"-6000#SCR'D.CPLG.				
1	1"-6000#SCR'D.CPLG.				
	SCH.40 A53-B SMLS.				
A 1	4" 7'-0" 2-BE				
B 1	4" 10'-13/4" 2-BE				
	TOTAL = 17'-2"				

COLOR CODE RED

NOTES

WT EACH   PAINT ▬
HEAT TR ▬

R.T. 5%	PRE-HEAT —	HYD-TEST —	HELIARC-R —	CODE B31.3	JOB NO	SPEC. 300	DATE 1-4	DET F.W.	CHK	GROUP NO	INV NO	
REQ'D ONE	PC MK MK-4"-P-1967-6		REF DWG P-1967		P O NO	REQ'D ONE		SHEET NO				SHEET NO

Fig. A-8 Spool (MK-4"-P-1967-6)

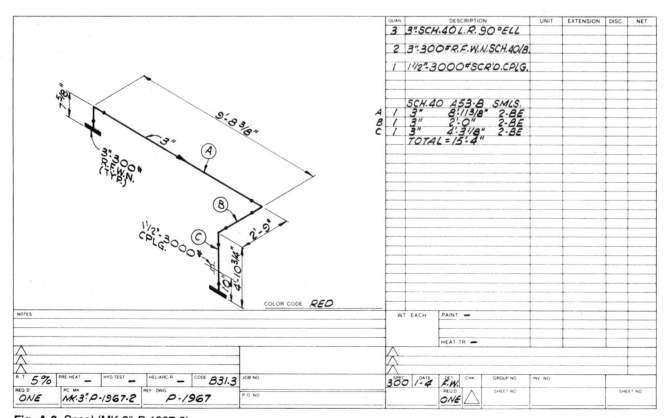

QUAN	DESCRIPTION	UNIT	EXTENSION	DISC.	NET
3	3"SCH.40 L.R. 90°ELL				
2	3"-300#R.F.W.N. SCH.40/B				
1	1 1/2"-3000#SCR'D.CPLG.				
	SCH.40 A53-B SMLS.				
A 1	3" 8'-11 3/8" 2-BE				
B 1	3" 2'-0" 2-BE				
C 1	3" 4'-3 1/8" 2-BE				
	TOTAL = 15'-4"				

COLOR CODE RED

NOTES

WT EACH   PAINT ▬
HEAT TR ▬

R.T. 5%	PRE-HEAT —	HYD-TEST —	HELIARC-R —	CODE B31.3	JOB NO	SPEC. 300	DATE 1-4	DET F.W.	CHK	GROUP NO	INV NO	
REQ'D ONE	PC MK MK-3"-P-1967-2		REF DWG P-1967		P O NO	REQ'D ONE		SHEET NO				SHEET NO

Fig. A-9 Spool (MK-3"-P-1967-2)

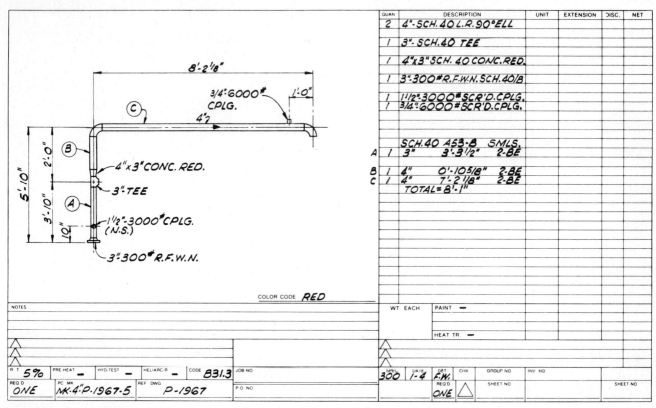

QUAN	DESCRIPTION	UNIT	EXTENSION	DISC.	NET
2	4"·SCH. 40 L.R. 90°ELL				
1	3"·SCH. 40 TEE				
1	4"x3" SCH. 40 CONC. RED.				
1	3"·300#R.F.W.N. SCH. 40/B				
1	1 1/2"·3000# SCR'D. CPLG.				
1	3/4"·6000# SCR'D. CPLG.				
	SCH. 40 A53·B SMLS.				
A 1	3" 3'·3 1/2" 2·BE				
B 1	4" 0'·10 5/8" 2·BE				
C 1	4" 7'·2 1/8" 2·BE				
	TOTAL = 8'·1"				

COLOR CODE **RED**

NOTES

WT EACH · PAINT ▬

HEAT TR ▬

R T 5%	PRE·HEAT ▬	HYD·TEST ▬	HELIARC·R ▬	CODE B31.3	JOB NO
REQ'D ONE	PC MK MK·4"·P·1967·5		REF DWG P·1967		P O NO

SPEC 300	DATE 1·4	DET F.W.	CHK	GROUP NO	INV NO
REQ'D ONE		SHEET NO			SHEET NO

Fig. A-10 Spool (MK-4"-P-1967-5)

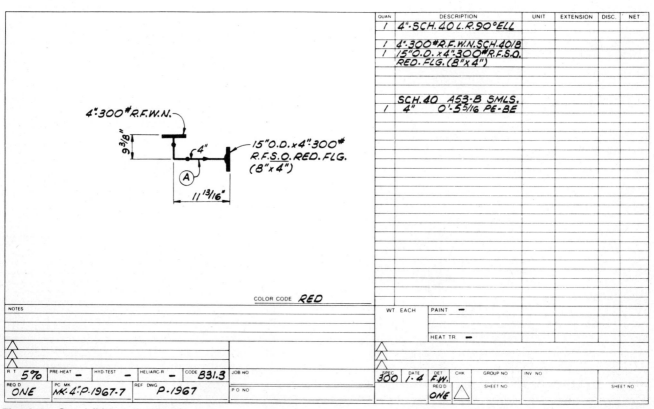

QUAN	DESCRIPTION	UNIT	EXTENSION	DISC.	NET
1	4"·SCH. 40 L.R. 90°ELL				
1	4"·300# R.F.W.N. SCH. 40/B				
1	15"O.D. x 4"·300# R.F.S.O. RED. FLG. (8"x 4")				
	SCH. 40 A53·B SMLS.				
1	4" 0'·5 5/16 PE·BE				

COLOR CODE **RED**

NOTES

WT EACH · PAINT ▬

HEAT TR ▬

R T 5%	PRE·HEAT ▬	HYD·TEST ▬	HELIARC·R ▬	CODE B31.3	JOB NO
REQ'D ONE	PC MK MK·4"·P·1967·7		REF DWG P·1967		P O NO

SPEC 300	DATE 1·4	DET F.W.	CHK	GROUP NO	INV NO
REQ'D ONE		SHEET NO			SHEET NO

Fig. A-11 Spool (MK-4"-P-1967-7)

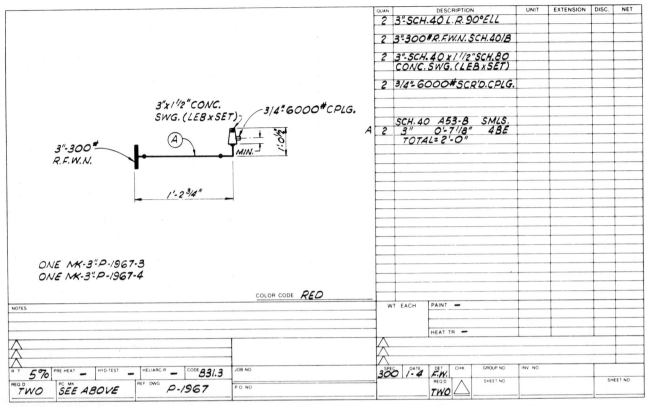

QUAN	DESCRIPTION	UNIT	EXTENSION	DISC.	NET
2	3" SCH. 40 L.R. 90° ELL				
2	3" 300# R.F.W.N. SCH.40/B				
2	3" SCH. 40 x 1 1/2" SCH. 80 CONC. SWG. (LEB x SET)				
2	3/4" 6000# SCR'D. CPLG.				
A 2	SCH. 40 A53-B SMLS. 3" 0'-7 1/8" 4 BE TOTAL = 2'-0"				

COLOR CODE	RED	
NOTES	WT EACH	PAINT —
		HEAT TR —
R T 5%	PRE-HEAT —	HYD-TEST —	HELIARC-R —	CODE 831.3	JOB NO	SPEC 300	DATE 1-4	DET F.W.	CHK	GROUP NO	INV NO
REQ D TWO	PC MK SEE ABOVE	REF DWG P-1967	P O NO		REQ'D TWO	SHEET NO			SHEET NO

Fig. A-12 Spools (MK-3″-P-1967-3 and (MK-3″-P-1967-4)

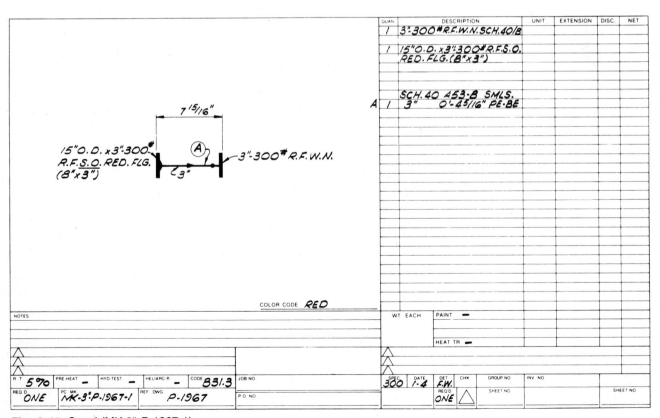

QUAN	DESCRIPTION	UNIT	EXTENSION	DISC.	NET
1	3" 300# R.F.W.N. SCH. 40/B				
1	15" O.D. x 3" 300# R.F.S.O. RED. FLG. (B" x 3")				
A 1	SCH. 40 A53-B SMLS. 3" 0'-4 5/16" PE-BE				

COLOR CODE	RED	
NOTES	WT EACH	PAINT —
		HEAT TR —
R T 5%	PRE-HEAT —	HYD-TEST —	HELIARC-R —	CODE 831.3	JOB NO	SPEC 300	DATE 1-4	DET F.W.	CHK	GROUP NO	INV NO
REQ D ONE	PC MK MK-3"-P-1967-1	REF DWG P-1967	P O NO		REQ'D ONE	SHEET NO			SHEET NO

Fig. A-13 Spool (MK-3″-P-1967-1)

Appendix B

The piping plans included in Appendix A should be accompanied by specifications such as those shown here in Appendix B. Although this is not a complete set of specifications, it is representative of the type of specifications that might be needed or used in the design and construction of a piping installation.

Various companies have their own system of writing specifications, but generally such specifications follow the same basic format as shown here. The purpose of such specifications is to provide additional information needed to design a pipe to fit any specific situation.

The student should study this set of specifications in relation to the plans in Appendix A.

	ISSUE DATE	SHT. NO.
GENERAL SPECIFICATION PIPING DESIGN AND INSTALLATION	April 1974	1 of 13
	General Spec. 5.1	

1. <u>SCOPE</u>

   This specification shall govern the mechanical design, material, installation and testing requirement for pipe, valves and fittings used in piping systems. This specification is not intended to govern the fabrication and inspection of piping systems.

2. <u>CODES AND REQUIREMENTS</u>

   2.1 All piping components, systems and design shall as a minimum, meet the requirements of ANSI B31.3, Petroleum Refinery Piping Code, latest edition, excepting only where specifically provided otherwise.

   2.2 Any local or national British code or regulation having the effect of law shall take precedence over these specifications.

3. <u>DESIGN</u>

   3.1 <u>General</u>

   3.1.1 All materials installed shall be new, clean and free from rust, pits and defects.

   3.1.2 No cast iron, ductile iron, malleable iron, aluminum plastic or copper bearing alloy shall be used in hydrocarbon service.

   3.1.3 Branch connections with a branch to header diameter ratio greater than 0.8 shall be made using weld tees. Reinforcing pads shall be used for diameter ratios in the range of 0.8 to 0.5. All branch connections less than 0.50 and in size 1-1/2" and under shall be made with items such as weldolets. As a minimum, all branch connections shall be designed and reinforced to equal the pressure rating at design temperature of the header. For specific requirements see individual piping classification charts.

   3.1.4 Piping shall be adequately supported for weight of piping water full, attached unsupported equipment, wind and seismic loading and where it becomes necessary to dismantel piping components for maintenance.

REV 0	

	ISSUE DATE	SHT. NO.
GENERAL SPECIFICATION PIPING DESIGN AND INSTALLATION	April 1974	2 of 13
	General Spec. 5.1	

3.1.5   No lines shall be run on floors, walkways, or skid grating, which would result in a tripping hazard.

3.1.6   All piping systems shall be provided with means of draining low points and venting high points. Such vents and drains shall meet all requirements and specifications for the systems to which they connect.

3.1.7   Instrument air, fuel gas and utility air takeoffs from a header shall be taken from the top side with a block valve at the header.

3.1.8   Temporary strainers shall be provided in all pump suction lines.

3.1.9   Where reduction is required at horizontal pump suction, use eccentric reducers with flat side up.

3.1.10   Any safety or relief valve discharging into a closed system shall be installed so that discharge piping slopes into header. If not possible, a 3/4 inch unvalved drain line shall be installed from the lowest point in the discharge piping to a safe area.

3.1.11   Any safety valve discharging to the atmosphere shall be equipped with riser. A 1/2 inch drain hole shall be provided in the lowest point of the riser. If the drain hole would permit a flame, developing from release of combustible material, that would impinge on equipment, the drain shall be piped to a safe location.

3.1.12   Forged fittings for low temperature service (-20°F and lower) shall meet ASTM A-594 specifications.

3.1.13   When specified for low temperature service (-20°F and lower) bolting materials shall meet ASTM A-320 requirements.

3.1.14   Any block valve installed between a safety or relief valve and the equipment which it protects shall be car sealed open or locked unless local codes or standards require otherwise.

REV 0

Pipe Drafting

GENERAL SPECIFICATION
PIPING DESIGN AND INSTALLATION

ISSUE DATE	SHT. NO.
April 1974	3 of 13

General Spec. 5.1

3.1.15   Any block valve installed between a safety or relief valve and the system into which it discharges shall be car sealed open or locked, unless local codes or standards require otherwise.

3.1.16   Drain and vent connections for all services 600 lb. and higher shall be double valved with a block valve adjacent to the pressure source and a throttling type valve on the immediate downstream side.

3.1.17   Long radius butt weld elbows shall be used wherever possible. Short radius ells may be used only where space requirements so dictate.

3.1.18   Lines containing valves shall be designed to prevent excessive stress concentrations on valves due to environmental or operating conditions. All piping shall be so arranged and supported as to impose no undue stress or transmit any undue force or vibration to any piece of equipment to which it attaches.

3.1.19   Expansion loops shall be used in preference to expansion joints where necessary due to thermal expansion or contraction. Cold springing may be used within the limits of B31.3. Anchors and guides shall be provided as required.

3.1.20   Line-size check valves shall be installed in the discharge piping from centrifugal pumps, compressors and other similar installations.

3.1.21   Lines subject to freezing shall be heat traced and insulated.

3.1.22   Mitred bends shall not be used.

3.2   Pipe

3.2.1   Process piping supplied under this specification shall conform to designations ASTM A-106 Grade B Seamless or API 5L Grade B Seamless. Low temperature piping (-20°F and below) shall meet ASTM A-333 specifications.

REV 0

3.2.2　Pipe wall thickness for all services governed by B31.3 shall be calculated by means of the formulas therein using a Y factor of 0.4.

3.2.3　Pipe with nominal size of 1-1/4, 2-1/2, 3-1/2, 5 and 7 inch shall not be used except as necessary to connect equipment. Where necessary the size shall be changed immediately adjacent to equipment.

3.2.4　Pipe 1-1/2 and smaller on process lines shall be Schedule 80 minimum thickness through 600 lb. services, Schedule 160 minimum for 900 through 1,500 lb. services and double extra strong minimum thickness for 2,500 lb. services. If piping is screwed the thread depth will be considered and wall thickness increased accordingly.

3.2.5　Minimum pipe size in any process service shall be 1/2 inch nominal diameter. Pipe headers and/or runs shall be 3/4 inch minimum size.

3.2.6　Corrosion allowances for piping shall be 0.063 inches for all sizes and service unless otherwise specified.

3.2.7　Piping joints in all services governed by this specification shall be butt welded for nominal sizes 2 inch and larger.

3.2.8　Piping joints for nominal sizes 1-1/2 inch and smaller shall be as follows: Unless noted otherwise on specific piping class charts:

a. Socket-welded in 600 pound services and above.
b. Screwed in 150 pound and 300 pound services.

3.2.9　Where use of screwed connections cannot be avoided in 600 pound services and higher, then joints shall be made up dry and seal welded excepting only where such welding is deemed injurious to equipment or instruments.

| rev1 | Revision Paragraph 3.2.3 | |
| REV 0 | | |

GENERAL SPECIFICATION PIPING DESIGN AND INSTALLATION	ISSUE DATE	SHT. NO.
	April 1974	5 of 13
	General Spec. 5.1	

3.3 <u>Valves</u>

3.3.1   Valves of whatever type and kind shall meet all requirements of this specification and shall be production models of manufacturers regularly engaged in valve manufacture.

3.3.2   Valves shall, as a minimum, meet all the requirements of the specified service class as set forth in Sub-section 3.6 of this specification.

3.3.3   Substitution of cast iron, ductile iron, bronze, brass or plastic bodied valves shall be permissible only when no other suitable alternative exists and only when so approved by the Engineer.

3.3.4   Valve sizes 2 inch and larger shall be flanged in all services and ratings.

3.3.5   Valve sizes 1-1/2 inch and smaller shall have socket weld ends as specified on individual piping class charts nominally rated process and utility service will be screwed for 150 pound and 300 pound rated services.

3.3.6   Block valves shall generally be installed at the inlets and outlets of all equipment, systems and control devices for purposes of safety and for maintenance.

3.3.7   Block valves shall generally be line size valves in accordance with the appropriate piping classification.

3.3.8   Block valves may be an approved type of gate, ball or non-lubricated plug valve where permissible in accordance with Sub-section 3.6 of this specification.

3.3.9   Bypasses around control valve stations shall generally be globe valves of the same nominal body size as the control valve.

3.3.10   All valves regardless of type and size will be located and oriented with due consideration for operability and maintenance.

REV 0

	ISSUE DATE	SHT. NO.
GENERAL SPECIFICATION PIPING DESIGN AND INSTALLATION	April 1974	6 of 13
	General Spec. 5.1	

3.3.11 Manually operated gate valves shall generally be equipped with bevel-gear operators as follows:

Specification	Size
L2	10" and Larger
L3	8" and Larger
L4	6" and Larger
L5 & L6	4" and Larger
L7	2" and Larger

3.3.12 Face-face dimensions for flanged end valves shall correspond to the requirements of ANSI B16.10, latest edition, where possible.

3.4 Flanges

3.4.1 All flanges 24 inch and smaller shall be furnished in strict accordance with ANSI B16.5, latest edition, or equivalent British Standards.

3.4.2 All flanges 26 inch and larger shall be in accordance with B31.3 and Manufacturers Standard Practice MSS-SP44, latest edition.

3.4.3 Flanges shall be raised face in all ANSI ratings through 600 lb. and ring type joint in all ANSI services 900 lb. and higher.

3.4.4 Weld neck flanges bored to the inside diameter of the pipe to which they are to be welded are preferred wherever space permits. Slip on flanges may be used only where space requirements dictate.

3.4.5 Flanges shall be installed with bolt holes straddling normal horizontal and vertical centerlines.

3.4.6 Flange material shall be ASTM A-105, or A-181, or an approved substitute.

3.4.7 Where flat face flanges are required for attachment to equipment these shall be carbon steel flanges with the raised face machined off. Raised face steel flanges shall not be bolted to cast iron or aluminum flanges.

REV 0

3.4.8   Forged Flanges for low temperature services (-20°F and lower) shall meet ASTM A350 specifications.

3.5   Threaded and Instrument Piping

     3.5.1   Screwed fittings shall be rated 3,000 lb. WOG for SCH 80 pipe and 6000 lb WOG when used for all pipe heavier than SCH 80 unless otherwise specified herein.

     3.5.2   Threaded piping shall be SCH 80 minimum wall thickness, unless otherwise specified herein.

     3.5.3   Use of threaded piping is limited to sizes 1-1/2 inch and smaller, unless otherwise specified herein.

     3.5.4   Use of close nipples and half couplings is prohibited.

     3.5.5   Screwed plugs shall be made from solid steel round or hex bar stock.

3.6   Line Numbering Procedure

     3.6.1   Line numbering shall be as follows:

         a.   The first part will designate line size.
         b.   The second part will designate service.
         c.   The third part will designate a consecutive number for identification purposes.
         d.   The fourth will designate the specification to be used.

         Example:

```
12"   -   HCL   -   101   -   L1
 |         |          |         |
 |         |          |         Specification
 |         |          Line Identification
 |         Service
Size
```

     3.6.2   Service designations shall be as follows:

REV 0		

	ISSUE DATE	SHT. NO.
GENERAL SPECIFICATION PIPING DESIGN AND INSTALLATION	April 1974	8 of 13
	General Spec. 5.1	

Service	Service Desig.	Spec. Nos.
Hydrocarbon Liquid	HCL	L2, L3, L5
Hydrocarbon Vapor	HCV	L2, L3, L5
Heat Transfer Medium	HO	L2, L3
Drain & Sewer	D	L2, L3, L5
Vent and Relief	F	L2, L3, L5
Diesel Fuel	DF	L2
Fuel Gas	FG	L2
Instrument Gas & Inert Gas	IG	L2
Blanket Gas	BG	L2
Air	A	L1
Water	W	L2, L3
Chemical Injection & Method	C	L2, L5
Glycol	GL	GL2, GL3
Hydrocarbon Liquid (-20F to -50)	HCL-LT	LT2, LT3
Hydrocarbon Vapor (-20F to -50)	HCV-LT	LT2, LT3

3.6.3   Pressure ratings shall be designated as follows:

125 lb.(A.N.S.I.)	L1
150 lb.(A.N.S.I.)	L2
300 lb.(A.N.S.I.)	L3
900 lb.(A.N.S.I.)	L5

4. INSTALLATION

4.1 Flanged Joint Assembly

Fabricator shall be responsible for proper assembly of flanged joints in accordance with the following procedure:

a.   Clean the finished flange surfaces thoroughly by approved means;

b.   Inspect surfaces for defects such as scratches or dents;

c.   Align surfaces in accordance with Sub-section 6.6 of this section of these Specifications;

d.   Install gasket carefully to accomplish centering without damage to the gasket;

e.   Install all flange bolts finger tight;

f.   Tighten one bolt initially to not over half the necessary value to seat the gasket and proceed to a second bolt adjacent to the diametrical opposite of the first. Proceed in like manner until all bolts have been partially tightened;

REV 0

g.   Repeat the same pattern of sequential bolting to obtain sufficient bolt stress to seat the gasket;

4.2   Threading

4.2.1   Threads shall be concentric with the outside barrel of the pipe and in accordance with ANSI Standard B2.1.

4.2.2   Threads shall be chased where necessary if impared by weld splatter or heat distortion due to the welding process.

4.2.3   Threaded joints for seal welding shall meet requirements of this section and be made up dry.

4.3   Handling

4.3.1   Coating

Any piping joints, fittings, flanges, spools, assemblies, etc., which are or may be exposed to the atmosphere for a prolonged period of time during or subsequent to fabrication and/or erection must be protected with a coat of rust preventative.

4.3.2   Protection of Openings

All piping spools or assemblies covered by 4.3.1 above shall be further protected as follows:

rev1	Deletes Paragraph 4.1, h and i	
REV 0		

    a.   Flanges shall be blanked with bolted or steel strapped wood or metal covers no smaller than the flange O.D.

    b.   Plain or beveled ends shall be closed with metal or plastic covers to protect the inside of the pipe.

    c.   Threaded connections shall be closed with a steel pipe plug or a thread protector.

4.4   Pipe Shoes and Anchors

4.4.1   Pipe Shoes and anchors shall be of the type and kind required by the drawings.

4.4.2   Standard structural shapes shall be used wherever possible.

4.4.3   Structural steel shall be in accordance with ASTM Specification A36, latest edition, or an approved substitute.

4.4.4   Structural materials shall be of new stock reasonably free from mill scale and rust and shall be reasonably straight and free from deformation.

4.4.5   Shoes or anchors to be attached to piping shall be welded in place prior to any stress relieving operations.

4.4.6   Edges exposed by cutting, shearing, etc., shall be left free of burrs.

4.5   Stress Relieving

Stress relieving shall be carried out where required in strict accordance with the code.

The method and equipment used in stress relieving must be acceptable to the Engineer.

4.6   Pressure Testing

4.6.1   Piping systems shall be pressure tested "in place" to the maximum extent practical.

4.6.2   All systems shall be reviewed to ensure that vents, drains, blind flanges, etc., are installed as necessary whether shown on drawings or not.

REV 0

	ISSUE DATE	SHT. NO.
GENERAL SPECIFICATION PIPING DESIGN AND INSTALLATION	April 1974	11 of 13
	General Spec. 5.1	

4.6.3   Closed piping systems shall generally be pressure tested in strict accordance with the applicable Code.

4.6.4   Open or low pressure systems may be air and soap bubble tested to 50 psig when acceptable to Engineer.

4.6.5   Positive provisions must be made to protect pressure sensitive equipment and instruments such as gauges, controllers, relief valves, etc., during testing.

4.6.6   All equipment, controls, instruments, etc., used in conducting pressure tests shall be in good repair and of a kind and type acceptable to the Engineer.

4.6.7   All pressure tests must be carried to a successful conclusion.

4.6.8   After testing is finished, piping systems shall be flushed with test fluid until the inside of the lines is thoroughly cleaned. After testing and flushing all lines shall be completely drained.

4.6.9   Pressure tests shall be conducted over a one-hour minimum time period and recorded by a proven instrument.

4.6.10 Water for testing shall be clear, treated water free of objectionable minerals, salts and organic materials.

4.6.11 Where freezing is a danger during testing, the testing medium shall include glycol or methanol in sufficient concentration to eliminate the possibility.

4.6.12 Until completion of pressure testing, all piping joints (including flanged, threaded or welded) shall be left unpainted and uninsulated a minimum distance of 2 inches on either side of the joint.

4.6.13 Control valves shall be open or have the plugs removed during the pressure test.

REV 0

	ISSUE DATE	SHT. NO.
GENERAL SPECIFICATION PIPING DESIGN AND INSTALLATION	April 1974	12 of 13
	General Spec. 5.1	

4.6.14   Upon conclusion of a pressure test all equipment and instrumentation either isolated or removed for purposes of the test shall be reinstalled or otherwise made ready for start up and all test equipment shall be disconnected and removed from the system in question.

4.6.15   The following equipment shall not be subjected to piping test pressures:

   a.   Pumps, turbines and compressors.
   b.   Explosion discs, safety valves, flame arrestors and filters.
   c.   Any equipment which does not have a specified test pressure.
   d.   Equipment not subject to pressure testing in conjunction with associated piping shall be isolated from the piping by spectacle blinds, blind flanges, plugs or caps. Valves shall not be relied upon in lieu of such isolating devices.

4.6.16   Instrument take-off piping up the the first block valve shall be tested with the piping or equipment to which it is connected. Testing of remainder of lead line up to instrument can also be done at the same time provided instruments are blocked off from source of pressure and vented.

REV 0

	ISSUE DATE	SHEET NO.
45° TO 90° BRANCH CONNECTION DETAILS **FOR — INTERNAL PRESSURE PIPING SYSTEMS**	April 1974	13 of 13
	STD DWG.	E 3-3

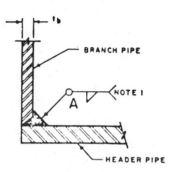

WELDED REINFORCEMENT OR NON-REINFORCED
90° BRANCH CONNECTIONS

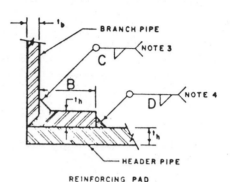

REINFORCING PAD
FOR 90° BRANCH CONNECTIONS

DRILL & TAP PAD FOR 1/8" IPS TEST HOLE
(DO NOT PLUG)
IF REINFORCING PAD IS ADDED IN TWO
PIECES, A HOLE IS REQUIRED THRU
EACH PIECE.

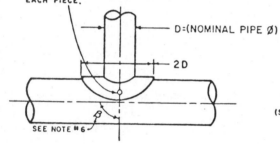

D = (NOMINAL PIPE Ø)

2D

SEE NOTE #6

LEGEND

T = STRAIGHT OR REDUCING TEE

PAD = REINFORCING PAD (NOTE 6)

WOL = WELDOLET

$*$ { SOL = SOCKOLET

TOL = THREDOLET

NOL = NIPPLE-O-LET

(SEE NOTE 5)

NOTES:

BRANCH REINFORCEMENT TABLES ON INDIVIDUAL PIPING SPECIFICATIONS INDICATE THE NECESSARY REINFORCEMENT FOR BRANCH CONNECTIONS.

1. TABULATED DIMENSIONS LESS THAN 1" INDICATE THE SIZE OF FILLET WELD "A".

2. TABULATED DIMENSIONS 1" OR LARGER INDICATE WIDTH "B" OF A REINFORCING PAD. REINFORCING PADS SHALL BE CUT FROM PIPE OF SAME WALL THICKNESS AND MATERIAL AS HEADER AND MAY BE SAME SIZE OR NEXT PIPE SIZE LARGER FORMED TO FIT.

3. FILLET WELD "C" SHALL BE THE GREATER OF 3/8" OR BRANCH THICKNESS (t_b) MINUS CORROSION ALLOWANCE.

4. FILLET WELD "D" SHALL BE EQUAL TO THE PAD THICKNESS (t_h) TIMES 0.707.

5. ($*$) INDICATED ON CHARTS REQUIRES PROPER SELECTION FOR ONE OF THE FOLLOWING CONDITIONS.

   SOL = WELDED PIPING TIEING INTO HEADER ($1\frac{1}{2}$" OR LESS)

   TOL = VENT OR INSTRUMENT CONNECTIONS ($1\frac{1}{2}$" OR LESS) WHERE ALLOWED BY PIPING SPEC.

   NOL = PIPING WITH VALVE AT HEADER ($1\frac{1}{2}$" OR LESS)

6. WHERE "PAD" IS SPECIFIED IN PIPING SPEC USE 2D FOR PAD O.D. ON 90° BRANCH CONNECTIONS. ON LATERAL TYPE BRANCH CONNECTIONS UP TO AND INCLUDING 45° USE 2D/SIN. β FOR PAD WIDTH. FOR PAD WIDTH, D IS ALWAYS MEASURED ALONG THE MAJOR AXIS.

REV 0				

	ISSUE DATE	SHEET NO.
PIPING - GATE VALVES	MAY 1974	E4-3-001
DESCRIPTION AND EQUIVALENT LIST		

INDEX	SPECIFICATIONS AND STANDARDS(WHERE APPLICABLE)

GATE VALVES

VG-1 THROUGH 20   BRASS OR BRONZE VALVES

VG-21 THROUGH 29   IRON VALVES

VG-40 THROUGH 80   STEEL VALVES, FLANGED, SCREWED OR SOCKET WELD

VG-100C THROUGH 120C   LOW TEMP VALVES (-20 TO -50) FLANGED, SCREWED, OR SOCKET WELD

2" OR SMALLER, FLANGED, SCREWED OR SOCKET WELD

VG-20	1/2"-2" - 150 LBS. BRONZE BODY, SCREWED CONNECTIONS INSIDE SCREW RISING STEM, SOLID WEDGE, (CRANE NO. 431 OR EQUAL)
VG-30	1-1/2" AND SMALLER 600 LB FS SW OUTSIDE SCREW AND YOKE, BOLTED BONNET, CRANE, 3607XU OR EQUAL
VG-70	1500# STEEL SOCKET WELD GATE VALVE 11-13% CHROME TRIM, STELLITE FACED SEATS, OS&Y BOLTED BONNET, SOLID WEDGE VOGT SW-1033 1-1/2" AND SMALLER.
VG-40	1-1/2" AND SMALLER 600 LB. FS SCREWED, OUT-SIDE SCREW AND YOKE, BOLTED BONNET, CRANE 3607XU OR EQUAL.
VG-100C	800# ASTM A-350-LF-1 FS GATE VALVE, SOCKET WELD, 18-8 SS TRIM, OS&Y BOLTED BONNET SOLID WEDGE, SERVICE: LPG LIQUID AND VAPOR UP TO -50°F. (SMITH C800A SW)

IRON BODY VALVES FLANGED

VG-21	3" AND LARGER - 125 LBS. IRON BODY, BRONZE TRIMMED FLAT FACE FLANGED CONNECTIONS, OUTSIDE SCREW AND YOKE, BOLTED BONNET, SOLID WEDGE. (CRANE NO. 456- 1/2 OR EQUAL).

STEEL VALVES FLANGED

VG-41	2" 150 LB. CS FLANGED, OS&Y, BOLTED BONNET, CRANE #47XU OR EQUAL.
VG-42	3" THROUGH 8" 150 LB. CS FLANGED, OS&Y BOLTED BONNET, HANDWHEEL OPERATED, CRANE #47 X OR EQUAL.
VG-43	10" AND LARGER 150 LB. CS FLANGED, OX&Y, BOLTED BONNET, BEVELGEAR OPERATED, CRANE #47 X OR EQUAL.
VG-44	1 1/2" & SMALLER 150 LB. CS FLANGED, OS&Y BOLTED BONNET, SMITH FIG. 815 OR EQUAL.
VG-50	2" 300 LB. CS FLANGED, OS&Y, BOLTED BONNET, CRANE #33 XU OR EQUAL.
VG-51	3" THROUGH 6" 300 LB. CS FLANGED, OS&Y BOLTED BONNET, HANDWHEEL OPERATED, CRANE #33 X OR EQUAL.
VG-52	3" THROUGH 6" 300 LB. CS FLANGED, OS&Y, BOLTED BONNET, BEVELGEAR OPERATED, CRANE #33 X OR EQUAL
VG-53	1 1/2" AND SMALLER 300 LBS. CS FLANGED, OS&Y BOLTED BONNET, SMITH FIG 830 OR EQUAL.
VG-61	600# STEEL FLANGED 1/4" RF GATE VALVE, 11-13% CHROME TRIM, RENEWABLE SEAT RINGS, OS&Y BOLTED BONNET, SOLID WEDGE PACIFIC 650-1 OR EQUAL.
VG-71	900# RF STEEL FLANGED GATE VALVE 11-13% CHROME TRIM OS&Y BOLTED BONNET, SOLID WEDGE, PACIFIC 950-1 OR EQUAL.
VG-101C	150# ASTM A-352-LCB RF GATE VALVE 18-8 CR-NI RENEWABLE SEAT RINGS, OS&Y, BOLTED BONNET SOLID WEDGE. SERVICE: LPG LIQUID AND VAPOR -50°F (PACIFIC 150-10 MODIFIED OR EQUAL). 1-1/2 " THRU 8"
VG-102C	300# ASTM A-352-LCB RF GATE VALVE 18-8 CR-NI RENEWABLE SEAT RINGS, OS&Y, BOLTED BONNET, SOLID WEDGE. SERVICE: LPG LIQUID AND VAPOR -50° F (PACIFIC 350-10 MODIFIED OR EQUAL). 1-1/2" THRU
VG-103C	150# ASTM A-352-LCB RF GATE VALVE 18-8 CR-NI RENEWABLE SEAT RINGS, OS&Y, BOLTED BONNET SOLID WEDGE, BEVELGEAR OPERATED. SERVICE: LPG LIQUID AND VAPOR -50°F (PACIFIC 150-10 MODIFIED OR EQUAL). 10" THRU 24"
VG-104C	300# ASTM A-352-LCB RF GATE VALVE 18-8 CR-NI RENEWABLE SEAT RINGS, OS&Y, BOLTED BONNET, SOLID WEDGE BEVELGEAR OPERATED, SERVICE: LPG LIQUID AND VAPOR- 50°F (PACIFIC 350-10 MODIFIED OR EQUAL). 8" THRU 24"

NATIONAL TANK COMPANY

REV	1	NOTED		REV					

	ISSUE DATE	SHEET NO.
PIPING - GLOBE VALVES **DESCRIPTION AND EQUIVALENT LIST**	MAY 1974	E4-3-002

INDEX	SPECIFICATIONS AND STANDARDS (WHERE APPLICABLE)

VGL-1 THROUGH VGL-20   BRASS OR BRONZE VALVES
 	
VGL-21 THROUGH VGL-29   IRON VALVES
 	
VGL-40 THROUGH VGL-80   STEEL VALVES FLANGED, SCREWED OR
         SOCKET WELD
 	
VGL-101C THROUGH VGL-120C   LOW TEMP. VALVES (-20 TO -50) FLANGED,
         SCREWED OR SOCKET WELD	

2" OR SMALLER SCREWED VALVES	2" OR LARGER STEEL VALVES
VGL-20 1/2"-2" 150 LB. BRONZE BODY, I.S.R.S., SCREWED CONNECTIONS, COMPOSITION DISC, INTEGRAL SEAT. (CRANE NO. 7 OR EQUAL).	**VGL-42** 2" THROUGH 8" 150 LB CS FLANGED, OS&Y, BOLTED BONNET, CRANE #143XR OR EQUAL.
⚠ **VGL-30** 1" AND 1-1/2" 600 LB FS SOCKET WELD, INSIDE SCREW BOLTED BONNET, CRANE 3624 XW OR EQUAL.	**VGL-50** 2" THROUGH 6" 300 LB. CS FLANGED, OS&Y, BOLTED BONNET, CRANE #151U OR EQUAL.
VGL-40 3/4" AND SMALLER 600 LB. FS SCREWED, INSIDE SCREW, UNION BONNET, CRANE 3620XW OR EQUAL.	**VGL-61** 600# STEEL, 1/4" RF FLANGED GLOBE VALVE, 11-13% CHROME TRIM, OS&Y BOLTED BONNET, PLUG TYPE DISC., RENEWABLE SEAT PACIFIC 660-1 OR EQUAL.
VGL-41 1" AND 1-1/2" 600 LB FS SCREWED, INSIDE SCREW, BOLTED BONNET, CRANE #3624XW OR EQUAL.	**VGL-71** 900# STEEL 1/4" RF FLANGED GLOBE VALVE 11-13% CHROME TRIM OS&Y BOLTED BONNET PLUG TYPE DISC RENEWABLE SEAT (POWELL 9031 F E OR EQUAL).
VGL-70 1500# STEEL SOCKET WELD GLOBE VALVE, 11-13% CHROME STELLITE FACED SEATS OS&Y BOLTED BONNET PLUG TYPE DISC., INTEGRAL SEAT-VOGT SW 1023 THRU SW-1027 OR EQUAL.	**VGL-101C** 150# ASTM-4352-LCB 1-1/2" GLOBE VALVE, 18-8 SS TRIM, OS&Y, BOLTED BONNET, PLUG TYPE DISC., RENEWABLE SEAT, SERVICE LPG LIQUID AND VAPOR TO-50°F. (PACIFIC 160-10 MODIFIED OR EQUAL). ⚠
	VGL-102C 300# ASTM-A352-LCB RF GLOBE VALVE, 18-8 SS TRIM, OS&Y, BOLTED BONNET, PLUG TYPE DISC., RENEWABLE SEAT. SERVICE LP OR LIQUID AND VAPOR TO -50°F. (PACIFIC 360-10 MODIFIED OR EQUAL).

REV	1	NOTED	REV			

	ISSUE DATE	SHEET NO.
PIPING - CHECK VALVES DESCRIPTION AND EQUIVALENT LIST	MAY 1974	E-4-3-003

INDEX	SPECIFICATIONS AND STANDARDS (WHERE APPLICABLE)
VC-1　THROUGH 20　BRASS OR BRONZE VALVES VC-21　THROUGH 29　IRON VALVES VC-40　THROUGH 80　STEEL VALVES FLANGED, SCREWED, OR SOCKET WELD VC-101C THROUGH 120C　LOW TEMP VALVES (-20 TO -50) FLANGED, SCREWED OR SOCKET WELD	

2" OR SMALLER FLANGED, SCREWED OR SOCKET WELD		STEEL VALVES FLANGED	
VC-20	1/2" - 2" - 200 LB. BRONZE BODY, SCREWED CONNECTION, BRONZE DISC., SWING CHECK (CRANE NO. 36 OR EQUAL).	VC-41	2" THROUGH 4 150 LB. 410 SS FLANGED, MISSION DUO-CHECK #15 EPF OR EQUAL. 11-13% CR. TRIM, METAL SEATS
⚠ VC-30	1-1/2" AND SMALLER 600 LB. CS SOCKET WELD, SWING CHECK CRANE #3694X OR EQUAL.	VC-42	6" THROUGH 36" 150 LB. CS FLANGED, MISSION DUO-CHECK #(15 SPF-111) OR EQUAL 11-13%CR. TRIM, METAL SEATS
VC-40	1-1/2" AND SMALLER 600 LB. CS SCREWED, SWING CHECK, CRANE #3694X OR EQUAL.	VC-50	2" THROUGH 4" 300 LB. 410 SS FLANGED, MISSION DUO-CHECK 30 EPF OR EQUAL. 11-13% CR. TRIM, METAL SEATS
VC-70	1500# STEEL, SOCKET WELD, PISTON CHECK VALVE, BOLTED CAP, INTEGRAL SEAT, HORIZONTAL OR VERTICAL, 316 SS TRIM, STELLITE FACED SEATS (EDWARD #1038 Y OR EQUAL).	VC-51	6" THROUGH 24" 300 LB. RF CS FLANGED MISSION DUO-CHECK 30 SPF OR EQUAL. 11-13% CR TRIM METAL SEATS
VC-101C	800# ASTM-A 350-LF-1 FS CHECK VALVE, 18-8 SS TRIM, BOLTED CAP, INTEGRAL SEAT, HORIZONTAL LIFT TYPE, SERVICE: LPG LIQUID AND VAPOR UP TO -50°F. (SMITH C-80 SW MODIFIED OR EQUAL).	VC-61	600# 410 SS FLANGED 1/4" RF CHECK VALVE METAL SEAT 11-13% CHROME TRIM MISSION 60 EPF OR EQUAL.
		VC-62	600# STEEL FLANGED 1/4" RF CHECK VALVE, METAL SEAT 11-13% CHROME TRIM MISSION 60 SPF OR EQUAL.
IRON BODY VALVES FLANGED		VC-71 -20°-600°F	900# 410 SS FLANGED 1/4" RF CHECK VALVE METAL SEAT 11-13% CHROME TRIM (MISSION 90 EPF OR EQUAL).
VC-21	3" AND LARGER - 125 LB. FF FLANGED, BOLTED CAP, RENEWABLE SEAT, SWING TYPE, I.B.B.T. (CRANE NO. 373 OR EQUAL).	VC-72 -20°-600°	900# STEEL FLANGED 1/4" RF CHECK VALVE METAL SEAT 11-13% CHROME TRIM (MISSION 90 SPF-111 OR EQUAL).
		VC-102C	150# ASTM A352-LCB RF CHECK VALVE, 316 SS TRIM, METAL SEAL. SERVICE: LPG LIQUID AND VAPOR TO -50°F. MISSION 15 GPF-XXX.
		VC-103C	300# ASTM A352-LCB RF CHECK VALVE, 316 SS TRIM METAL SEAL. SERVICE: LPG LIQUID AND VAPOR TO -50°F. MISSION 30 GPF-XXX.

REV	1	NOTED	REV			

SHEET NO.	ISSUE DATE	SERVICE		TEMP RANGE	CORR. ALL.	RATING	CLASS
E4-2-001	MAY 1974			-20/200°F	0.063"	150 LB	L-1
CODE ANSI B31.3		INSTRUMENT & UTILITY AIR SYSTEM				200 PSIG	

PIPE AND FITTINGS / VALVES

PIPE

SIZE	THICKNESS & MAT'L
2" & SMLR.	SCH 80 A 120 SMLS GALV.
3"-6"	SCH.40 A-53B SMLS

SCREWED FITTINGS (GALV)

3000 LB F.S. A181 GR.I	XS A105 GRII	SCH 80 API 5L GRB SMLS
90 DEG. ELL	FULL COUPLING	NIPPLE TBE A181
45 DEG. ELL		CONC. SWG. NIP TBE
STRAIGHT TEE		ECC SWG NIP TBE
REDUCING TEE		
REDUCER		
CAP		
UNION (INTEGRAL SEATS)		

BUTT-WELDING FITTINGS
ANSI B16.9 A234 GR B

Size	90° LR		90° SR		45° LR		TEE		CAP		CONC RED		ECC RED	
3-6"	STD.	WT.	STD.	WT.	STD.	WT.	STD.	WT.	STD.	WT.	STD.	WT.	STD.	WT.

GATE

SIZE	DESCRIPTION	CODE
2" & SMLR	150 LB SCR'D. BRONZE GATE VALVE	VG-20
3"-6"	125 LB FF IRON BODY BRONZE TRIM GATE VALVE	VG-21
1 1/2"	150 LB RF GATE VALVE (500°F)	VG-44 ⚠

MECHANICAL JOINTS

SIZE	DESCRIPTION
	FLANGE ANSI B16.5 A181 GR. 1
1 1/2"- 2"	150 LB RF THREADED FLANGE
3"-6"	150 LB RF WN FLANGE STD. BORE

⚠

BOLTS & GASKETS

STUD BOLTS PER ASTM A 93 GR B7 W/TWO HEX NUTS PER A 194 GR 2H CAD. PL. GASKETS (150 LB FF) 1/16" COMP D. ASBESTOS-FULL FACE PER ANSI B16.21.

PLUG

SIZE	DESCRIPTION	CODE

BALL

SIZE	DESCRIPTION	CODE

MISCELLANEOUS ITEMS

SIZE	DESCRIPTION

GLOBE

SIZE	DESCRIPTION	CODE
2" & SMLR	150 LB SCR'D BRONZE GLOBE VALVE	VGL-20
3"-6"	125 LB FF IRON BODY GLOBE VALVE	VGL-21

CHECK

SIZE	DESCRIPTION	CODE
2" & SMLR	200 LB SCR'D. BRONZE CHECK VALVE	VC-20
3"-6"	125 LB FF IRON BODY SWING TYPE CHECK VALVE	VC-21

BRANCH REINFORCEMENT

	1" & SMLR	1-1/2	2	3	4	6
1" & SMLR	T					
1-1/2	T	T				
2	*	*	T			
3	*	*	T	T		
4	*	*	PAD	T	T	
6	*	*	WOL	PAD	T	T

SMLR 1-1/2 2 3 4 6
BRANCH SIZES
HEADER / SIZES

⚠ NOTE:

1. REFER TO STD DWG E3-3 FOR PAD SIZES

   *   = SEE STD. DWG. E3-3
   WOL = WELDOLET
   PAD = REINFORCING PAD
   T   = TEE

OTHER VALVES

SIZE	DESCRIPTION	CODE

NOTES:

PRESS-TEMP RATING

TEMP °F	PRESS PSIG MAX	TEMP °F	PRESS PSIG MAX
-20/100	200		

MAX. HYDROSTATIC TEST 300 PSIG LIMITED BY DESIGN.

NATIONAL TANK COMPANY

REV.	1	NOTED	REV.

SHEET NO.	ISSUE DATE	SERVICE		TEMP RANGE	CORR. ALL.	RATING	CLASS
E4-2-002	MAY 1974	**S E R V I C E**		-20/550°F	0.063"	150 LB RF CARB. STL	L-2
C O D E ANSI B31.3		**PRIMARY SERVICE** (SEE NOTE NO. 2 BELOW)					

P I P E   A N D   F I T T I N G S   V A L V E S

PIPE

SIZE	THICKNESS & MAT'L
1-1/2" & SMLR	XS A-106 GR. B. SMLS
2"-24"	STD. WT. A 106 GR. B SMLS
30"	STD. WT. API-5L GR. B SAW (JOINT FACTOR 1.0)

SCREWED FITTINGS

3000 LB FS A105 GR II	XS A105 GR II	XS A106 GR B
90 DEG. ELL	THREDOLET	NIPPLE TBE SMLS/A-105-11
45 DEG. ELL	ELBOLET, REG.	CONC. SWG. NIP TBE
STRAIGHT TEE	ELBOLET, LAT.	CONC. SWG. NIP TBE/TSE
REDUCING TEE	NIPOLET	ECC. SWG. NIP TBE
REDUCER	SOLID ROUND HEAD PLUG	ECC. SWG. NIP BLE/TSE
UNION (INTEGRAL SEATS)		
FULL COUPLING		

BUTT - WELDING FITTINGS
ANSI B16.9 A234 GR. B

Size	90° LR	90° SR	45° LR	TEE	CAP	CONC. RED	ECC. RED
2-30"	STD. WT.	STD. WT.	STD. WT.	STD. WT.	STD. WT.	STD. WT	STD. WT.

M E C H A N I C A L   J O I N T S

SIZE	DESCRIPTION
⚠ 1 1/2"-24"	FLANGES PER ANSI B16.5 ASTM A 181 GR. II
	150 LB RF WN FLANGE STD. WT. BORE (SEE NOTE #2)
2"-24"	300 LB RF WN ORIFICE FLANGE, STD. WT. BORE
30"	150 LB RF BS 3293 STD. BORE

BOLTS & GASKETS

STUD BOLTS PER ASTM A193 GR. B7 W/TWO HEX NUTS PER A194, GR. 2H CAD. PL. 2" THRU 30" GASKETS (150 lb RF) 1/16" COMP'D ASBESTO-FLAT RING PER ANSI B16.21 (DITTO 300 LB RF)
FOR HOT OIL SERVICE SEE NOTE #1

MISCELLANEOUS ITEMS

SIZE	DESCRIPTION

BRANCH REINFORCEMENT

NOTE:
1. REFER TO STD. DWG. E5-3 FOR PAD SIZES
   * = SEE STD. DWG. E3-3
   WOL = WELDOLET
   PAD = REINFORCING PAD
   T = TEE

GATE

SIZE	DESCRIPTION	CODE
1-1/2 & SMLR	600 LB SCR'D GATE VALVE (850°F)	VG-40
2"	150 LB RF GATE VALVE (850°F)	VG-41
3"-6"	150 LB RF GATE VALVE (850°F)	VG-42
10"-24"	150 LB RF GATE VALVE (850°F)	VG-43
1 1/2"	150 LB RF GATE VALVE (500°F) (SEE NOTE #2)	VG-44 ⚠

PLUG

SIZE	DESCRIPTION	CODE

BALL

SIZE	DESCRIPTION	CODE

GLOBE

SIZE	DESCRIPTION	CODE
3/4" & SMLR	600 LB SCR'D GLOBE VALVE (850°F)	VGL-40
1" & 1-1/2"	600 LB SCR'D GLOBE VALVE (850°F)	VGL-41
2"-12"	150 LB RF GLOBE VALVE (500°F)	VGL-42

CHECK

SIZE	DESCRIPTION	CODE
1-1/2 & SMLR	600 LB SCR'D SWG CHECK VALVE (850°F)	VC-40
2"-4"	150 LB RF DUO-CHECK VALVE (600°F)	VC-41
6"-36"	150 LB RF DUO-CHECK VALVE (600°F)	VC-42

OTHER VALVES

SIZE	DESCRIPTION	CODE
	SAMPLE VALVE	
1/4-1/2"	3,000 SCR'D NEEDLE VALVE (800°F)	VN-3

NOTES:

⚠ 1. 2" THRU 24" GASKETS 1/8" SPIRAL WOUND WITH 316 SS METAL WINDING & ASBESTOS PER API-601 (HOT OIL SERVICE ONLY)

⚠ 2. 1 1/2" WN FLG'S USED FOR INST. HOOK-UPS OR AGAINST FITTINGS TO BE 150 LB RF WN-STD BORE

2. THIS CLASS IS FOR THE FOLLOWING SERVICES:

HCL –	DF –	W –
HCV –	FG –	C –
HO –	IG –	F –
D –	BG –	

PRESS - TEMP RATING

TEMP °F	PRESS PSIG MAX	TEMP °F	PRESS PSIG MAX
-20/100	275		
150	255		
200	240		
250	225		
300	210		
400	180		
500	165		
600	130		

NATIONAL TANK COMPANY

REV.	1	NOTED	REV	

INDEX